Optimal Control with Engineering Applications

Hans P. Geering

Optimal Control with Engineering Applications

With 12 Figures

 Springer

Hans P. Geering, Ph.D.
Professor of Automatic Control and Mechatronics
Measurement and Control Laboratory

Department of Mechanical and Process Engineering
ETH Zurich
Sonneggstrasse 3
CH-8092 Zurich, Switzerland

Library of Congress Control Number: 2007920933

ISBN 978-3-540-69437-3 Springer Berlin Heidelberg New York

Springer is a part of Springer Science+Business Media

springer.com

© Springer-Verlag Berlin Heidelberg 2007

Typesetting: Camera ready by author
Production: LE-TeX Jelonek, Schmidt & Vöckler GbR, Leipzig
Cover design: eStudioCalamar S.L., F. Steinen-Broo, Girona, Spain

SPIN 11880127   7/3100/YL - 5 4 3 2 1 0   Printed on acid-free paper

Foreword

This book is based on the lecture material for a one-semester senior-year undergraduate or first-year graduate course in optimal control which I have taught at the Swiss Federal Institute of Technology (ETH Zurich) for more than twenty years. The students taking this course are mostly students in mechanical engineering and electrical engineering taking a major in control. But there also are students in computer science and mathematics taking this course for credit.

The only prerequisites for this book are: The reader should be familiar with dynamics in general and with the state space description of dynamic systems in particular. Furthermore, the reader should have a fairly sound understanding of differential calculus.

The text mainly covers the design of open-loop optimal controls with the help of Pontryagin's Minimum Principle, the conversion of optimal open-loop to optimal closed-loop controls, and the direct design of optimal closed-loop optimal controls using the Hamilton-Jacobi-Bellman theory.

In theses areas, the text also covers two special topics which are not usually found in textbooks: the extension of optimal control theory to matrix-valued performance criteria and Lukes' method for the iterative design of approximatively optimal controllers.

Furthermore, an introduction to the phantastic, but incredibly intricate field of differential games is given. The only reason for doing this lies in the fact that the differential games theory has (exactly) one simple application, namely the LQ differential game. It can be solved completely and it has a very attractive connection to the H_∞ method for the design of robust linear time-invariant controllers for linear time-invariant plants. — This route is the easiest entry into H_∞ theory. And I believe that every student majoring in control should become an expert in H_∞ control design, too.

The book contains a rather large variety of optimal control problems. Many of these problems are solved completely and in detail in the body of the text. Additional problems are given as exercises at the end of the chapters. The solutions to all of these exercises are sketched in the Solution section at the end of the book.

Acknowledgements

First, my thanks go to Michael Athans for elucidating me on the background
of optimal control in the first semester of my graduate studies at M.I.T. and
for allowing me to teach his course in my third year while he was on sabbatical
leave.

I am very grateful that Stephan A. R. Hepner pushed me from teaching the
geometric version of Pontryagin's Minimum Principle along the lines of [2],
[20], and [14] (which almost no student understood because it is so easy, but
requires 3D vision) to teaching the variational approach as presented in this
text (which almost every student understands because it is so easy and does
not require any 3D vision).

I am indebted to Lorenz M. Schumann for his contributions to the material
on the Hamilton-Jacobi-Bellman theory and to Roberto Cirillo for explaining
Lukes' method to me.

Furthermore, a large number of persons have supported me over the years. I
cannot mention all of them here. But certainly, I appreciate the continuous
support by Gabriel A. Dondi, Florian Herzog, Simon T. Keel, Christoph
M. Schär, Esfandiar Shafai, and Oliver Tanner over many years in all aspects
of my course on optimal control. — Last but not least, I like to mention my
secretary Brigitte Rohrbach who has always eagle-eyed my texts for errors
and silly faults.

Finally, I thank my wife Rosmarie for not killing me or doing any other
harm to me during the very intensive phase of turning this manuscript into
a printable form.

Hans P. Geering

Fall 2006

Contents

List of Symbols

Independent Variables

t	time
t_a, t_b	initial time, final time
t_1, t_2	times in (t_a, t_b),
	e.g., starting end ending times of a singular arc
τ	a special time in $[t_a, t_b]$

Vectors and Vector Signals

$u(t)$	control vector, $u(t) \in \Omega \subseteq R^m$
$x(t)$	state vector, $x(t) \in R^n$
$y(t)$	output vector, $y(t) \in R^p$
$y_d(t)$	desired output vector, $y_d(t) \in R^p$
$\lambda(t)$	costate vector, $\lambda(t) \in R^n$,
	i.e., vector of Lagrange multipliers
q	additive part of $\lambda(t_b) = \nabla_x K(x(t_b)) + q$
	which is involved in the transversality condition
λ_a, λ_b	vectors of Lagrange multipliers
$\mu_0, \ldots, \mu_{\ell-1}, \mu_\ell(t)$	scalar Lagrange multipliers

Sets

$\Omega \subseteq R^m$	control constraint
$\Omega_u \subseteq R^{m_u}$, $\Omega_v \subseteq R^{m_v}$	control constraints in a differential game
$\Omega_x(t) \subseteq R^n$	state constraint
$S \subseteq R^n$	target set for the final state $x(t_b)$
$T(S, x) \subseteq R^n$	tangent cone of the target set S at x
$T^*(S, x) \subseteq R^n$	normal cone of the target set S at x
$T(\Omega, u) \subseteq R^m$	tangent cone of the constraint set Ω at u
$T^*(\Omega, u) \subseteq R^m$	normal cone of the constraint set Ω at u

Integers

i, j, k, ℓ	indices
m	dimension of the control vector
n	dimension of the state and the costate vector
p	dimension of an output vector
λ_0	scalar Lagrange multiplier for J,
	1 in the regular case, 0 in a singular case

Functions

$f(.)$	function in a static optimization problem
$f(x, u, t)$	right-hand side of the state differential equation
$g(.)$, $G(.)$	define equality or inequality side-constraints
$h(.)$, $g(.)$	switching function for the control and offset function
	in a singular optimal control problem
$H(x, u, \lambda, \lambda_0, t)$	Hamiltonian function
$J(u)$	cost functional
$\mathcal{J}(x, t)$	optimal cost-to-go function
$L(x, u, t)$	integrand of the cost functional
$K(x, t_b)$	final state penalty term
$A(t)$, $B(t)$, $C(t)$, $D(t)$	system matrices of a linear time-varying system
F, $Q(t)$, $R(t)$, $N(t)$	penalty matrices in a quadratic cost functional
$G(t)$	state-feedback gain matrix
$K(t)$	solution of the matrix Riccati differential equation
	in an LQ regulator problem
$P(t)$	observer gain matrix
$Q(t)$, $R(t)$	noise intensity matrices in a stochastic system
$\Sigma(t)$	state error covariance matrix
$\kappa(.)$	support function of a set

Operators

$\frac{d}{dt}$, $\dot{}$	total derivative with respect to the time t
$\mathrm{E}\{\ldots\}$	expectation operator
$[\ldots]^{\mathrm{T}}$, T	taking the transpose of a matrix
U	adding a matrix to its transpose
$\dfrac{\partial f}{\partial x}$	Jacobi matrix of the vector function f
	with respect to the vector argument x
$\nabla_x L$	gradient of the scalar function L with respect to x,

$$\nabla_x L = \left(\frac{\partial L}{\partial x}\right)^{\mathrm{T}}$$

1 Introduction

1.1 Problem Statements

In this book, we consider two kinds of dynamic optimization problems: optimal control problems and differential game problems.

In an optimal control problem for a dynamic system, the task is finding an admissible control trajectory $u : [t_a, t_b] \to \Omega \subseteq R^m$ generating the corresponding state trajectory $x : [t_a, t_b] \to R^n$ such that the cost functional $J(u)$ is minimized.

In a zero-sum differential game problem, one player chooses the admissible control trajectory $u : [t_a, t_b] \to \Omega_u \subseteq R^{m_u}$ and another player chooses the admissible control trajectory $v : [t_a, t_b] \to \Omega_v \subseteq R^{m_v}$. These choices generate the corresponding state trajectory $x : [t_a, t_b] \to R^n$. The player choosing u wants to minimize the cost functional $J(u, v)$, while the player choosing v wants to maximize the same cost functional.

1.1.1 The Optimal Control Problem

We only consider optimal control problems where the initial time t_a and the initial state $x(t_a) = x_a$ are specified. Hence, the most general optimal control problem can be formulated as follows:

Optimal Control Problem:
Find an admissible optimal control $u : [t_a, t_b] \to \Omega \subseteq R^m$ such that the dynamic system described by the differential equation

$$\dot{x}(t) = f(x(t), u(t), t)$$

is transferred from the initial state

$$x(t_a) = x_a$$

into an admissible final state

$$x(t_b) \in S \subseteq R^n \ ,$$

and such that the corresponding state trajectory $x(.)$ satisfies the state constraint

$$x(t) \in \Omega_x(t) \subseteq R^n$$

at all times $t \in [t_a, t_b]$, and such that the cost functional

$$J(u) = K(x(t_b), t_b) + \int_{t_a}^{t_b} L(x(t), u(t), t)\, dt$$

is minimized.

Remarks:

1) Depending upon the type of the optimal control problem, the final time t_b is fixed or free (i.e., to be optimized).

2) If there is a nontrivial control constraint (i.e., $\Omega \neq R^m$), the admissible set $\Omega \subset R^m$ is time-invariant, closed, and convex.

3) If there is a nontrivial state constraint (i.e., $\Omega_x(t) \neq R^n$), the admissible set $\Omega_x(t) \subset R^n$ is closed and convex at all times $t \in [t_a, t_b]$.

4) Differentiability: The functions f, K, and L are assumed to be at least once continuously differentiable with respect to all of their arguments.

1.1.2 The Differential Game Problem

We only consider zero-sum differential game problems, where the initial time t_a and the initial state $x(t_a) = x_a$ are specified and where there is no state constraint. Hence, the most general zero-sum differential game problem can be formulated as follows:

Differential Game Problem:
Find admissible optimal controls $u : [t_a, t_b] \to \Omega_u \subseteq R^{m_u}$ and $v : [t_a, t_b] \to \Omega_v \subseteq R^{m_v}$ such that the dynamic system described by the differential equation

$$\dot{x}(t) = f(x(t), u(t), v(t), t)$$

is transferred from the initial state

$$x(t_a) = x_a$$

to an admissible final state

$$x(t_b) \in S \subseteq R^n$$

and such that the cost functional

$$J(u) = K(x(t_b), t_b) + \int_{t_a}^{t_b} L(x(t), u(t), v(t), t)\, dt$$

is minimized with respect to u and maximized with respect to v.

Remarks:

1) Depending upon the type of the differential game problem, the final time t_b is fixed or free (i.e., to be optimized).

2) Depending upon the type of the differential game problem, it is specified whether the players are restricted to open-loop controls $u(t)$ and $v(t)$ or are allowed to use state-feedback controls $u(x(t), t)$ and $v(x(t), t)$.

3) If there are nontrivial control constraints, the admissible sets $\Omega_u \subset R^{m_u}$ and $\Omega_v \subset R^{m_v}$ are time-invariant, closed, and convex.

4) Differentiability: The functions f, K, and L are assumed to be at least once continuously differentiable with respect to all of their arguments.

1.2 Examples

In this section, several optimal control problems and differential game problems are sketched. The reader is encouraged to wonder about the following questions for each of the problems:

• Existence: Does the problem have an optimal solution?

• Uniqueness: Is the optimal solution unique?

• What are the main features of the optimal solution?

• Is it possible to obtain the optimal solution in the form of a state feedback control rather than as an open-loop control?

Problem 1: Time-optimal, friction-less, horizontal motion of a mass point

State variables:
$$x_1 = \text{position}$$
$$x_2 = \text{velocity}$$

control variable:
$$u = \text{acceleration}$$

subject to the constraint
$$u \in \Omega = [-a_{\max}, +a_{\max}] \ .$$

Find a piecewise continuous acceleration $u : [0, t_b] \rightarrow \Omega$, such that the dynamic system
$$\begin{bmatrix} \dot{x}_1(t) \\ \dot{x}_2(t) \end{bmatrix} = \begin{bmatrix} 0 & 1 \\ 0 & 0 \end{bmatrix} \begin{bmatrix} x_1(t) \\ x_2(t) \end{bmatrix} + \begin{bmatrix} 0 \\ 1 \end{bmatrix} u(t)$$

is transferred from the initial state
$$\begin{bmatrix} x_1(0) \\ x_2(0) \end{bmatrix} = \begin{bmatrix} s_a \\ v_a \end{bmatrix}$$

to the final state

$$\begin{bmatrix} x_1(t_b) \\ x_2(t_b) \end{bmatrix} = \begin{bmatrix} s_b \\ v_b \end{bmatrix}$$

in minimal time, i.e., such that the cost criterion

$$J(u) = t_b = \int_0^{t_b} dt$$

is minimized.

Remark: s_a, v_a, s_b, v_b, and a_{max} are fixed.

For obvious reasons, this problem is often named "time-optimal control of the double integrator". It is analyzed in detail in Chapter 2.1.4.

Problem 2: Time-optimal, horizontal motion of a mass with viscous friction

This problem is almost identical to Problem 1, except that the motion is no longer frictionless. Rather, there is a friction force which is proportional to the velocity of the mass.

Thus, the equation of motion (with $c > 0$) now is:

$$\begin{bmatrix} \dot{x}_1(t) \\ \dot{x}_2(t) \end{bmatrix} = \begin{bmatrix} 0 & 1 \\ 0 & -c \end{bmatrix} \begin{bmatrix} x_1(t) \\ x_2(t) \end{bmatrix} + \begin{bmatrix} 0 \\ 1 \end{bmatrix} u(t) \ .$$

Again, find a piecewise continuous acceleration $u : [0, t_b] \rightarrow [-a_{max}, a_{max}]$ such that the dynamic system is transferred from the given initial state to the required final state in minimal time.

In contrast to Problem 1, this problem may fail to have an optimal solution. Example: Starting from stand-still with $v_a = 0$, a final velocity $|v_b| > a_{max}/c$ cannot be reached.

Problem 3: Fuel-optimal, friction-less, horizontal motion of a mass point

State variables:

$$x_1 = \text{position}$$
$$x_2 = \text{velocity}$$

control variable:

$$u = \text{acceleration}$$

subject to the constraint

$$u \in \Omega = [-a_{max}, +a_{max}] \ .$$

Find a piecewise continuous acceleration $u : [0, t_b] \rightarrow \Omega$, such that the dynamic system

$$\begin{bmatrix} \dot{x}_1(t) \\ \dot{x}_2(t) \end{bmatrix} = \begin{bmatrix} 0 & 1 \\ 0 & 0 \end{bmatrix} \begin{bmatrix} x_1(t) \\ x_2(t) \end{bmatrix} + \begin{bmatrix} 0 \\ 1 \end{bmatrix} u(t)$$

is transferred from the initial state

$$\begin{bmatrix} x_1(0) \\ x_2(0) \end{bmatrix} = \begin{bmatrix} s_a \\ v_a \end{bmatrix}$$

to the final state

$$\begin{bmatrix} x_1(t_b) \\ x_2(t_b) \end{bmatrix} = \begin{bmatrix} s_b \\ v_b \end{bmatrix}$$

and such that the cost criterion

$$J(u) = \int_0^{t_b} |u(t)| \, dt$$

is minimized.

Remark: s_a, v_a, s_b, v_b, $a_{\max}$, and t_b are fixed.

This problem is often named "fuel-optimal control of the double integrator". The notion of fuel-optimality associated with this type of cost functional relates to the physical fact that in a rocket engine, the thrust produced by the engine is proportional to the rate of mass flow out of the exhaust nozzle. However, in this simple problem statement, the change of the total mass over time is neglected. — This problem is analyzed in detail in Chapter 2.1.5.

Problem 4: Fuel-optimal horizontal motion of a rocket

In this problem, the horizontal motion of a rocket is modeled in a more realistic way: Both the aerodynamic drag and the loss of mass due to thrusting are taken into consideration. State variables:

$$x_1 = \text{position}$$
$$x_2 = \text{velocity}$$
$$x_3 = \text{mass}$$

control variable:

$$u = \text{thrust force delivered by the engine}$$

subject to the constraint

$$u \in \Omega = [0, F_{\max}] \ .$$

The goal is minimizing the fuel consumption for a required mission, or equivalently, maximizing the mass of the rocket at the final time.

Thus, the optimal control problem can be formulated as follows:

Find a piecewise continuous thrust $u : [0, t_b] \to [0, F_{\max}]$ of the engine such that the dynamic system

$$\begin{bmatrix} \dot{x}_1(t) \\ \dot{x}_2(t) \\ \dot{x}_3(t) \end{bmatrix} = \begin{bmatrix} x_2(t) \\ \frac{1}{x_3(t)} \{u(t) - \frac{1}{2} A\rho c_w x_2^2(t)\} \\ -\alpha u(t) \end{bmatrix}$$

is transferred from the initial state

$$\begin{bmatrix} x_1(0) \\ x_2(0) \\ x_3(0) \end{bmatrix} = \begin{bmatrix} s_a \\ v_a \\ m_a \end{bmatrix}$$

to the (incompletely specified) final state

$$\begin{bmatrix} x_1(t_b) \\ x_2(t_b) \\ x_3(t_b) \end{bmatrix} = \begin{bmatrix} s_b \\ v_b \\ \text{free} \end{bmatrix}$$

and such that the equivalent cost functionals $J_1(u)$ and $J_2(u)$ are minimized:

$$J_1(u) = \int_0^{t_b} u(t)\, dt$$

$$J_2(u) = -x_3(t_b) \ .$$

Remark: s_a, v_a, m_a, s_b, v_b, $F_{\max}$, and t_b are fixed.

This problem is analyzed in detail in Chapter 2.6.3.

Problem 5: The LQ regulator problem

Find an unconstrained control $u : [t_a, t_b] \to R^m$ such that the linear time-varying dynamic system

$$\dot{x}(t) = A(t)x(t) + B(t)u(t)$$

is transferred from the initial state

$$x(t_a) = x_a$$

to an arbitrary final state

$$x(t_b) \in R^n$$

and such that the quadratic cost functional

$$J(u) = \frac{1}{2} x^{\mathrm{T}}(t_b) F x(t_b) + \frac{1}{2} \int_{t_a}^{t_b} \Big(x^{\mathrm{T}}(t) Q(t) x(t) + u^{\mathrm{T}}(t) R(t) u(t) \Big) dt$$

is minimized.

Remarks:

1) The final time t_b is fixed. The matrices F and $Q(t)$ are symmetric and positive-semidefinite and the matrix $R(t)$ is symmetric and positive-definite.

2) Since the cost functional is quadratic and the constraints are linear, automatically a linear solution results, i.e., the result will be a linear state feedback controller of the form $u(t) = -G(t)x(t)$ with the optimal time-varying controller gain matrix $G(t)$.

3) Usually, the LQ regulator is used in order to robustly stabilize a nonlinear dynamic system around a nominal trajectory:

Consider a nonlinear dynamic system for which a nominal trajectory has been designed for the time interval $[t_a, t_b]$:

$$\dot{X}_{\mathrm{nom}}(t) = f(X_{\mathrm{nom}}(t), U_{\mathrm{nom}}(t), t)$$
$$X_{\mathrm{nom}}(t_a) = X_a \ .$$

In reality, the true state vector $X(t)$ will deviate from the nominal state vector $X_{\mathrm{nom}}(t)$ due to unknown disturbances influencing the dynamic system. This can be described by

$$X(t) = X_{\mathrm{nom}}(t) + x(t) \ ,$$

where $x(t)$ denotes the state error which should be kept small by hopefully small control corrections $u(t)$, resulting in the control vector

$$U(t) = U_{\mathrm{nom}}(t) + u(t) \ .$$

If indeed the errors $x(t)$ and the control corrections can be kept small, the stabilizing controller can be designed by linearizing the nonlinear system around the nominal trajectory.

This leads to the LQ regulator problem which has been stated above. — The penalty matrices $Q(t)$ and $R(t)$ are used for shaping the compromise between keeping the state errors $x(t)$ and the control corrections $u(t)$, respectively, small during the whole mission. The penalty matrix F is an additional tool for influencing the state error at the final time t_b.

The LQ regulator problem is analyzed in Chapters 2.3.4 and 3.2.3. — For further details, the reader is referred to [1], [2], [16], and [25].

Problem 6: Goh's fishing problem

In the following simple economic problem, consider the number of fish $x(t)$ in an ocean and the catching rate of the fishing fleet $u(t)$ of catching fish per unit of time, which is limited by a maximal capacity, i.e., $0 \leq u(t) \leq U$. The goal is maximizing the total catch over a fixed time interval $[0, t_b]$.

The following reasonably realistic optimal control problem can be formulated:

Find a piecewise continuous catching rate $u : [0, t_b] \rightarrow [0, U]$, such that the fish population in the ocean satisfying the population dynamics

$$\dot{x}(t) = a\left(x(t) - \frac{x^2(t)}{b}\right) - u(t)$$

with the initial state

$$x(0) = x_a$$

and with the obvious state constraint

$$x(t) \geq 0 \qquad \text{for all } t \in [0, t_b]$$

is brought up or down to an arbitrary final state

$$x(t_b) \geq 0$$

and such that the total catch is maximized, i.e., such that the cost functional

$$J(u) = -\int_0^{t_b} u(t)\, dt$$

is minimized.

Remarks:

1) $a > 0$, $b > 0$; x_a, t_b, and U are fixed.

2) This problem nicely reveals that the solution of an optimal control problem always is "as bad" as the considered formulation of the optimal control problem. This optimal control problem lacks any sustainability aspect: Obviously, the fish will be extinct at the final time t_b, if this is feasible. (Think of whaling or raiding in business economics.)

3) This problem has been proposed (and solved) in [18]. An even more interesting extended problem has been treated in [19], where there is a predator-prey constellation involving fish and sea otters. The competing sea otters must not be hunted because they are protected by law.

Goh's fishing problem is analyzed in Chapter 2.6.2.

Problem 7: Slender beam with minimal weight

A slender horizontal beam of length L is rigidly clamped at the left end and free at the right end. There, it is loaded by a vertical force F. Its cross-section is rectangular with constant width b and variable height $h(\ell)$; $h(\ell) \geq 0$ for $0 \leq \ell \leq L$. Design the variable height of the beam, such that the vertical deflection $s(\ell)$ of the flexible beam at the right end is limited by $s(L) \leq \varepsilon$ and the weight of the beam is minimal.

Problem 8: Circular rope with minimal weight

An elastic rope with a variable but circular cross-section is suspended at the ceiling. Due to its own weight and a mass M which is appended at its lower end, the rope will suffer an elastic deformation. Its length in the undeformed state is L. For $0 \leq \ell \leq L$, design the variable radius $r(\ell)$ within the limits $0 \leq r(\ell) \leq R$ such that the appended mass M sinks by δ at most and such that the weight of the rope is minimal.

Problem 9: Optimal flying maneuver

An aircraft flies in a horizontal plane at a constant speed v. Its lateral acceleration can be controlled within certain limits. The goal is to fly over a reference point (target) in any direction and as soon as possible.

The problem is stated most easily in an earth-fixed coordinate system (see Fig. 1.1). For convenience, the reference point is chosen at $x = y = 0$. The limitation of the lateral acceleration is expressed in terms of a limited angular turning rate $u(t) = \dot\varphi(t)$ with $|u(t)| \leq 1$.

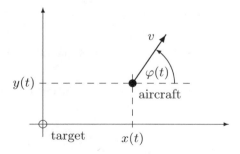

Fig. 1.1. Optimal flying maneuver described in earth-fixed coordinates.

Find a piecewise continuous turning rate $u : [0, t_b] \rightarrow [-1, 1]$ such that the dynamic system

$$\begin{bmatrix} \dot{x}(t) \\ \dot{y}(t) \\ \dot{\varphi}(t) \end{bmatrix} = \begin{bmatrix} v \cos \varphi(t) \\ v \sin \varphi(t) \\ u(t) \end{bmatrix}$$

is transferred from the initial state

$$\begin{bmatrix} x(0) \\ y(0) \\ \varphi(0) \end{bmatrix} = \begin{bmatrix} x_a \\ y_a \\ \varphi_a \end{bmatrix}$$

to the partially specified final state

$$\begin{bmatrix} x(t_b) \\ y(t_b) \\ \varphi(t_b) \end{bmatrix} = \begin{bmatrix} 0 \\ 0 \\ \text{free} \end{bmatrix}$$

and such that the cost functional

$$J(u) = \int_0^{t_b} dt$$

is minimized.

Alternatively, the problem can be stated in a coordinate system which is fixed to the body of the aircraft (see Fig. 1.2).

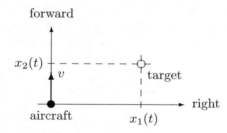

Fig. 1.2. Optimal flying maneuver described in body-fixed coordinates.

This leads to the following alternative formulation of the optimal control problem:

Find a piecewise continuous turning rate $u : [0, t_b] \rightarrow [-1, 1]$ such that the dynamic system

$$\begin{bmatrix} \dot{x}_1(t) \\ \dot{x}_2(t) \end{bmatrix} = \begin{bmatrix} x_2(t) u(t) \\ -v - x_1(t) u(t) \end{bmatrix}$$

is transferred from the initial state

$$\begin{bmatrix} x_1(0) \\ x_2(0) \end{bmatrix} = \begin{bmatrix} x_{1a} \\ x_{2a} \end{bmatrix} = \begin{bmatrix} -x_a \sin \varphi_a + y_a \cos \varphi_a \\ -x_a \cos \varphi_a - y_a \sin \varphi_a \end{bmatrix}$$

to the final state

$$\begin{bmatrix} x_1(t_b) \\ x_2(t_b) \end{bmatrix} = \begin{bmatrix} 0 \\ 0 \end{bmatrix}$$

and such that the cost functional

$$J(u) = \int_0^{t_b} dt$$

is minimized.

Problem 10: Time-optimal motion of a cylindrical robot

In this problem, the coordinated angular and radial motion of a cylindrical robot in an assembly task is considered (Fig. 1.3). A component should be grasped by the robot at the supply position and transported to the assembly position in minimal time.

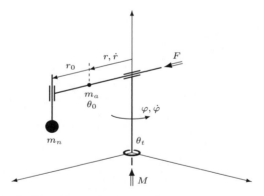

Fig. 1.3. Cylindrical robot with the angular motion φ and the radial motion r.

State variables:

$$x_1 = r = \text{radial position}$$
$$x_2 = \dot{r} = \text{radial velocity}$$
$$x_3 = \varphi = \text{angular position}$$
$$x_4 = \dot{\varphi} = \text{angular velocity}$$

control variables:

$$u_1 = F = \text{radial actuator force}$$

$$u_2 = M = \text{angular actuator torque}$$

subject to the constraints

$$|u_1| \leq F_{\max} \text{ and } |u_2| \leq M_{\max}, \text{ hence}$$

$$\Omega = [-F_{\max}, F_{\max}] \times [-M_{\max}, M_{\max}] \ .$$

The optimal control problem can be stated as follows:

Find a piecewise continuous $u : [0, t_b] \rightarrow \Omega$ such that the dynamic system

$$\begin{bmatrix} \dot{x}_1(t) \\ \dot{x}_2(t) \\ \dot{x}_3(t) \\ \dot{x}_4(t) \end{bmatrix} = \begin{bmatrix} x_2(t) \\ [u_1(t) + (m_a x_1(t) + m_n\{r_0 + x_1(t)\})x_4^2(t)]/(m_a + m_n) \\ x_4(t) \\ [u_2(t) - 2(m_a x_1(t) + m_n\{r_0 + x_1(t)\})x_2(t)x_4(t)]/\theta_{\text{tot}}(x_1(t)) \end{bmatrix}$$

where

$$\theta_{\text{tot}}(x_1(t)) = \theta_t + \theta_0 + m_a x_1^2(t) + m_n\{r_0 + x_1(t)\}^2$$

is transferred from the initial state

$$\begin{bmatrix} x_1(0) \\ x_2(0) \\ x_3(0) \\ x_4(0) \end{bmatrix} = \begin{bmatrix} r_a \\ 0 \\ \varphi_a \\ 0 \end{bmatrix}$$

to the final state

$$\begin{bmatrix} x_1(t_b) \\ x_2(t_b) \\ x_3(t_b) \\ x_4(t_b) \end{bmatrix} = \begin{bmatrix} r_b \\ 0 \\ \varphi_b \\ 0 \end{bmatrix}$$

and such that the cost functional

$$J(u) = \int_0^{t_b} dt$$

is minimized.

This problem has been solved in [15].

Problem 11: The LQ differential game problem

Find unconstrained controls $u : [t_a, t_b] \rightarrow R^{m_u}$ and $v : [t_a, t_b] \rightarrow R^{m_v}$ such that the dynamic system

$$\dot{x}(t) = A(t)x(t) + B_1(t)u(t) + B_2 v(t)$$

is transferred from the initial state

$$x(t_a) = x_a$$

to an arbitrary final state

$$x(t_b) \in R^n$$

at the fixed final time t_b and such that the quadratic cost functional

$$J(u, v) = \frac{1}{2}x^T(t_b)Fx(t_b)$$
$$+ \frac{1}{2} \int_{t_a}^{t_b} \left(x^T(t)Q(t)x(t) + u^T(t)u(t) - \gamma^2 v^T(t)v(t) \right) dt$$

is simultaneously minimized with respect to u and maximized with respect to v, when both of the players are allowed to use state-feedback control.

Remark: As in the LQ regulator problem, the penalty matrices F and $Q(t)$ are symmetric and positive-semidefinite.

This problem is analyzed in Chapter 4.2.

Problem 12: The homicidal chauffeur game

A car driver (denoted by "pursuer" P) and a pedestrian (denoted by "evader" E) move on an unconstrained horizontal plane. The pursuer tries to kill the evader by running him over. The game is over when the distance between the pursuer and the evader (both of them considered as points) diminishes to a critical value d. — The pursuer wants to minimize the final time t_b while the evader wants to maximize it.

The dynamics of the game are described most easily in an earth-fixed coordinate system (see Fig. 1.4).

State variables: x_p, y_p, φ_p, and x_e, y_e.

Control variables: $u \sim \dot{\varphi}_p$ ("constrained motion") and v_e ("simple motion").

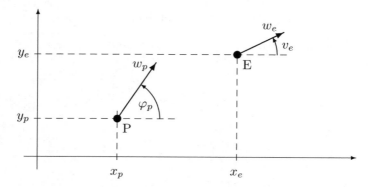

Fig. 1.4. The homicidal chauffeur game described in earth-fixed coordinates.

Equations of motion:

$$\dot{x}_p(t) = w_p \cos \varphi_p(t)$$

$$\dot{y}_p(t) = w_p \sin \varphi_p(t)$$

$$\dot{\varphi}_p(t) = \frac{w_p}{R} u(t) \qquad\qquad |u(t)| \le 1$$

$$\dot{x}_e(t) = w_e \cos v_e(t) \qquad\quad w_e < w_p$$

$$\dot{y}_e(t) = w_e \sin v_e(t)$$

Alternatively, the problem can be stated in a coordinate system which is fixed to the body of the car (see Fig. 1.5).

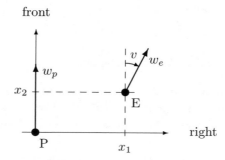

Fig. 1.5. The homicidal chauffeur game described in body-fixed coordinates.

This leads to the following alternative formulation of the differential game problem:

State variables: x_1 and x_2.

Control variables: $u \in [-1, +1]$ and $v \in [-\pi, \pi]$.

Using the coordinate transformation

$$x_1 = (x_e - x_p) \sin \varphi_p - (y_e - y_p) \cos \varphi_p$$

$$x_2 = (x_e - x_p) \cos \varphi_p + (y_e - y_p) \sin \varphi_p$$

$$v = \varphi_p - v_e \ ,$$

the following model of the dynamics in the body-fixed coordinate system is obtained:

$$\dot{x}_1(t) = \frac{w_p}{R} x_2(t) u(t) + w_e \sin v(t)$$

$$\dot{x}_2(t) = -\frac{w_p}{R} x_1(t) u(t) - w_p + w_e \cos v(t) \ .$$

Thus, the differential game problem can finally be stated in the following efficient form:

Find two state-feedback controllers $u(x_1, x_2) \mapsto [-1, +1]$ and $v(x_1, x_2) \mapsto [-\pi, +\pi]$ such that the dynamic system

$$\dot{x}_1(t) = \frac{w_p}{R} x_2(t) u(t) + w_e \sin v(t)$$

$$\dot{x}_2(t) = -\frac{w_p}{R} x_1(t) u(t) - w_p + w_e \cos v(t)$$

is transferred from the initial state

$$x_1(0) = x_{10}$$

$$x_2(0) = x_{20}$$

to a final state with

$$x_1^2(t_b) + x_2^2(t_b) \le d^2$$

and such that the cost functional

$$J(u, v) = t_b$$

is minimized with respect to $u(.)$ and maximized with respect to $v(.)$.

This problem has been stipulated and partially solved in [21]. The complete solution of the homicidal chauffeur problem has been derived in [28].

1.3 Static Optimization

In this section, some very basic facts of elementary calculus are recapitulated which are relevant for minimizing a continuously differentiable function of several variables, without or with side-constraints.

The goal of this text is to generalize these very simple necessary conditions for a constrained minimum of a function to the corresponding necessary conditions for the optimality of a solution of an optimal control problem. The generalization from constrained static optimization to optimal control is very straightforward, indeed. No "higher" mathematics is needed in order to derive the theorems stated in Chapter 2.

1.3.1 Unconstrained Static Optimization

Consider a scalar function of a single variable, $f : R \to R$. Assume that f is at least once continuously differentiable when discussing the first-order necessary condition for a minimum and at least k times continuously differentiable when discussing higher-order necessary or sufficient conditions.

The following conditions are *necessary* for a local minimum of the function $f(x)$ at x^o:

- $f'(x^o) = \dfrac{df(x^o)}{dx} = 0$

- $f^\ell(x^o) = \dfrac{d^\ell f(x^o)}{dx^\ell} = 0$ for $\ell = 1, \ldots, 2k-1$
  and $f^{2k}(x^o) \geq 0$ where $k = 1$, or 2, or, ...

The following conditions are *sufficient* for a local minimum of the function $f(x)$ at x^o:

- $f'(x^o) = \dfrac{df(x^o)}{dx} = 0$ and $f''(x^o) > 0$ or

- $f^\ell(x^o) = \dfrac{d^\ell f(x^o)}{dx^\ell} = 0$ for $\ell = 1, \ldots, 2k-1$
  and $f^{2k}(x^o) > 0$ for a finite integer number $k \geq 1$.

Nothing can be inferred from these conditions about the *existence* of a local or a global minimum of the function f!

If the range of admissible values x is restricted to a finite, closed, and bounded interval $\Omega = [a, b] \subset R$, the following conditions apply:

- If f is continuous, there exists at least one global minimum.

- Either the minimum lies at the left boundary a, and the lowest non-vanishing derivative is positive,

  or

  the minimum lies at the right boundary b, and the lowest non-vanishing derivative is negative,

  or

  the minimum lies in the interior of the interval, i.e., $a < x^o < b$, and the above-mentioned necessary and sufficient conditions of the unconstrained case apply.

Remark: For a function f of several variables, the first derivative f' generalizes to the Jacobian matrix $\frac{\partial f}{\partial x}$ as a row vector or to the gradient $\nabla_x f$ as a column vector,

$$\frac{\partial f}{\partial x} = \left[\frac{\partial f}{\partial x_1}, \ \cdots, \ \frac{\partial f}{\partial x_n} \right] \ , \qquad \nabla_x f = \left(\frac{\partial f}{\partial x} \right)^{\mathrm{T}} \ ,$$

the second derivative to the Hessian matrix

$$\frac{\partial^2 f}{\partial x^2} = \begin{bmatrix} \dfrac{\partial^2 f}{\partial x_1^2} & \cdots & \dfrac{\partial^2 f}{\partial x_1 \partial x_n} \\ \vdots & & \vdots \\ \dfrac{\partial^2 f}{\partial x_n \partial x_1} & \cdots & \dfrac{\partial^2 f}{\partial x_n^2} \end{bmatrix}$$

and its positive-semidefiniteness, etc.

1.3.2 Static Optimization under Constraints

For finding the minimum of a function f of several variables $x_1, \ldots, x_n$ under the constraints of the form $g_i(x_1, \ldots, x_n) = 0$ and/or $g_i(x_1, \ldots, x_n) \leq 0$, for $i = 1, \ldots, \ell$, the method of Lagrange multipliers is extremely helpful.

Instead of minimizing the function f with respect to the independent variables $x_1, \ldots, x_n$ over a constrained set (defined by the functions g_i), minimize the augmented function F with respect to its mutually completely independent variables $x_1, \ldots, x_n, \lambda_1, \ldots, \lambda_\ell$, where

$$F(x_1, \ldots, x_n, \lambda_1, \ldots, \lambda_\ell) = \lambda_0 f(x_1, \ldots, x_n) + \sum_{i=1}^{\ell} \lambda_i g_i(x_1, \ldots, x_n) \ .$$

Remarks:

- In shorthand, F can be written as $F(x, \lambda) = \lambda_0 f(x) + \lambda^{\mathrm{T}} g(x)$ with the vector arguments $x \in R^n$ and $\lambda \in R^\ell$.

- Concerning the constant λ_0, there are only two cases: it attains either the value 0 or 1.

  In the singular case, $\lambda_0 = 0$. In this case, the ℓ constraints uniquely determine the admissible vector x^o. Thus, the function f to be minimized is not relevant at all. Minimizing f is not the issue in this case! Nevertheless, minimizing the augmented function F still yields the correct solution.

  In the regular case, $\lambda_0 = 1$. The ℓ constraints define a nontrivial set of admissible vectors x, over which the function f is to be minimized.

- In the case of equality side constraints: since the variables $x_1, \ldots, x_n$, $\lambda_1, \ldots, \lambda_\ell$ are independent, the necessary conditions of a minimum of the augmented function F are

$$\frac{\partial F}{\partial x_i} = 0 \quad \text{for } i = 1,\ldots,n \qquad \text{and} \qquad \frac{\partial F}{\partial \lambda_j} = 0 \quad \text{for } j = 1,\ldots,\ell \,.$$

  Obviously, since F is linear in λ_j, the necessary condition $\frac{\partial F}{\partial \lambda_j} = 0$ simply returns the side constraint $g_i = 0$.

- For an inequality constraint $g_i(x) \leq 0$, two cases have to be distinguished: Either the minimum x^o lies in the interior of the set defined by this constraint, i.e., $g_i(x^o) < 0$. In this case, this constraint is irrelevant for the minimization of f because for all x in an infinitesimal neighborhood of x^o, the strict inequality holds; hence the corresponding Lagrange multiplier vanishes: $\lambda_i^o = 0$. This constraint is said to be inactive. — Or the minimum x^o lies at the boundary of the set defined by this constraint, i.e., $g_i(x^o) = 0$. This is almost the same as in the case of an equality constraint. Almost, but not quite: For the corresponding Lagrange multiplier, we get the necessary condition $\lambda_i^o \geq 0$. This is the so-called "Fritz-John" or "Kuhn-Tucker" condition [7]. This inequality constraint is said to be active.

Example 1: Minimize the function $f = x_1^2 - 4x_1 + x_2^2 + 4$ under the constraint $x_1 + x_2 = 0$.

Analysis for $\lambda_0 = 1$:

$$F(x_1, x_2, \lambda) = x_1^2 - 4x_1 + x_2^2 + 4 + \lambda x_1 + \lambda x_2$$

$$\frac{\partial F}{\partial x_1} = 2x_1 - 4 + \lambda = 0$$

$$\frac{\partial F}{\partial x_2} = 2x_2 + \lambda = 0$$

$$\frac{\partial F}{\partial \lambda} = x_1 + x_2 = 0 \,.$$

The optimal solution is:

$$x_1^o = 1$$
$$x_2^o = -1$$
$$\lambda^o = 2 \ .$$

Example 2: Minimize the function $f = x_1^2 + x_2^2$ under the constraints $1 - x_1 \leq 0$, $2 - 0.5x_1 - x_2 \leq 0$, and $x_1 + x_2 - 4 \leq 0$.

Analysis for $\lambda_0 = 1$:

$$F(x_1, x_2, \lambda_1, \lambda_2, \lambda_3) = x_1^2 + x_2^2$$
$$+ \lambda_1(1 - x_1) + \lambda_2(2 - 0.5x_1 - x_2) + \lambda_3(x_1 + x_2 - 4)$$

$$\frac{\partial F}{\partial x_1} = 2x_1 - \lambda_1 - 0.5\lambda_2 + \lambda_3 = 0$$

$$\frac{\partial F}{\partial x_2} = 2x_2 - \lambda_2 + \lambda_3 = 0$$

$$\frac{\partial F}{\partial \lambda_1} = 1 - x_1 \qquad \begin{cases} = 0 & \text{and} \ \ \lambda_1 \geq 0 \\ < 0 & \text{and} \ \ \lambda_1 = 0 \end{cases}$$

$$\frac{\partial F}{\partial \lambda_2} = 2 - 0.5x_1 - x_2 \begin{cases} = 0 & \text{and} \ \ \lambda_2 \geq 0 \\ < 0 & \text{and} \ \ \lambda_2 = 0 \end{cases}$$

$$\frac{\partial F}{\partial \lambda_3} = x_1 + x_2 - 4 \qquad \begin{cases} = 0 & \text{and} \ \ \lambda_3 \geq 0 \\ < 0 & \text{and} \ \ \lambda_3 = 0 \end{cases}$$

The optimal solution is:

$$x_1^o = 1$$
$$x_2^o = 1.5$$
$$\lambda_1^o = 0.5$$
$$\lambda_2^o = 3$$
$$\lambda_3^o = 0 \ .$$

The third constraint is inactive.

1.4 Exercises

1. In all of the optimal control problems stated in this chapter, the control
   constraint Ω is required to be a time-invariant set in the control space
   R^m.

   For the control of the forward motion of a car, the torque $T(t)$ delivered
   by the automotive engine is often considered as a control variable. It can
   be chosen freely between a minimal torque and a maximal torque, both
   of which are dependent upon the instantaneous engine speed $n(t)$. Thus,
   the torque limitation is described by

   $$T_{\min}(n(t)) \leq T(t) \leq T_{\max}(n(t)) \ .$$

   Since typically the engine speed is not constant, this constraint set for
   the torque $T(t)$ is not time-invariant.

   Define a new transformed control variable $u(t)$ for the engine torque such
   that the constraint set Ω for u becomes time-invariant.

2. In Chapter 1.2, ten optimal control problems are presented (Problems
   1–10). In Chapter 2, for didactic reasons, the general formulation of an
   optimal control problem given in Chapter 1.1 is divided into the categories
   A.1 and A.2, B.1 and B.2, C.1 and C.2, and D.1 and D.2. Furthermore, in
   Chapter 2.1.6, a special form of the cost functional is characterized which
   requests a special treatment.

   Classify all of the ten optimal control problems with respect to these
   characteristics.

3. Discuss the geometric aspects of the optimal solution of the constrained
   static optimization problem which is investigated in Example 1 in Chapter
   1.3.2.

4. Discuss the geometric aspects of the optimal solution of the constrained
   static optimization problem which is investigated in Example 2 in Chapter
   1.3.2.

5. Minimize the function $f(x, y) = 2x^2 + 17xy + 3y^2$ under the equality
   constraints $x - y = 2$ and $x^2 + y^2 = 4$.

2 Optimal Control

In this chapter, a set of necessary conditions for the optimality of a solution of
an optimal control problem is derived using the calculus of variations. This
set of necessary conditions is known by the name "Pontryagin's Minimum
Principle" [29]. Exploiting Pontryagin's Minimum Principle, several optimal
control problems are solved completely.

Solving an optimal control problem using Pontryagin's Minimum Principle
typically proceeds in the following (possibly iterative) steps:

- Formulate the optimal control problem.

- Existence: Determine whether the problem can have an optimal solution.

- Formulate all of the necessary conditions of Pontryagin's Minimum Principle.

- Globally minimize the Hamiltonian function H:
  $u^o(x^o(t), \lambda^o(t), \lambda_0^o, t) = \arg\min_{u \in \Omega} H(x^o(t), u, \lambda^o(t), \lambda_0^o, t)$ for all $t \in [t_a, t_b]$.

- Singularity: Determine whether the problem can have a singular solution.
  There are two scenarios for a singularity:
  a) $\lambda_0^o = 0$?
  b) $H \neq H(u)$ for $t \in [t_1, t_2]$? (See Chapter 2.6.)

- Solve the two-point boundary value problem for $x^o(\cdot)$ and $\lambda^o(\cdot)$.

- Eliminate locally optimal solutions which are not globally optimal.

- If possible, convert the resulting optimal open-loop control $u^o(t)$ into an
  optimal closed-loop control $u^o(x^o(t), t)$ using state feedback.

Of course, having the optimal control law in a feedback form rather than in
an open-loop form is advantageous in practice. In Chapter 3, a method is presented for designing closed-loop control laws directly in one step. It involves
solving the so-called Hamilton-Jacobi-Bellman partial differential equation.

For didactic reasons, the optimal control problem is categorized into several
types. In a problem of Type A, the final state is fixed: $x^o(t_b) = x_b$. In a
problem of Type C, the final state is free. In a problem of Type B, the final
state is constrained to lie in a specified target set S. — The Types A and

B are special cases of the Type C: For Type A: $S = \{x_b\}$ and for Type C: $S = R^n$.

The problem Type D generalizes the problem Type B to the case where there is an additional state constraint of the form $x^o(t) \in \Omega_x(t)$ at all times.

Furthermore, each of the four problem types is divided into two subtypes depending on whether the final time t_b is fixed or free (i.e., to be optimized).

2.1 Optimal Control Problems with a Fixed Final State

In this section, Pontryagin's Minimum Principle is derived for optimal control problems with a fixed final state (and no state constraints). The method of Lagrange multipliers and the calculus of variations are used.

Furthermore, two "classics" are presented in detail: the time-optimal and the fuel-optimal frictionless horizontal motion of a mass point.

2.1.1 The Optimal Control Problem of Type A

Statement of the optimal control problem:

Find a piecewise continuous control $u : [t_a, t_b] \to \Omega \subseteq R^m$, such that the constraints

$$x(t_a) = x_a$$
$$\dot{x}(t) = f(x(t), u(t), t) \qquad \text{for all } t \in [t_a, t_b]$$
$$x(t_b) = x_b$$

are satisfied and such that the cost functional

$$J(u) = K(t_b) + \int_{t_a}^{t_b} L(x(t), u(t), t)\, dt$$

is minimized;

Subproblem A.1: t_b is fixed (and $K(t_b) = 0$ is suitable),

Subproblem A.2: t_b is free ($t_b > t_a$).

Remark: t_a, $x_a \in R^n$, $x_b \in R^n$ are specified; $\Omega \subseteq R^m$ is time-invariant.

2.1.2 Pontryagin's Minimum Principle

Definition: Hamiltonian function $H : R^n \times \Omega \times R^n \times \{0,1\} \times [t_a, t_b] \to R$,

$$H(x(t), u(t), \lambda(t), \lambda_0, t) = \lambda_0 L(x(t), u(t), t) + \lambda^{\mathrm{T}}(t) f(x(t), u(t), t) .$$

Theorem A

If the control $u^o : [t_a, t_b] \to \Omega$ is optimal, then there exists a nontrivial vector

$$\begin{bmatrix} \lambda_0^o \\ \lambda^o(t_b) \end{bmatrix} \neq 0 \in R^{n+1} \quad \text{with } \lambda_0^o = \begin{cases} 1 & \text{in the regular case} \\ 0 & \text{in the singular case,} \end{cases}$$

such that the following conditions are satisfied:

a) $\dot{x}^o(t) = \nabla_\lambda H_{|o} = f(x^o(t), u^o(t), t)$

$\quad x^o(t_a) = x_a$

$\quad x^o(t_b) = x_b$

$\quad \dot{\lambda}^o(t) = -\nabla_x H_{|o} = -\lambda_0^o \nabla_x L(x^o(t), u^o(t), t) - \left[\dfrac{\partial f}{\partial x}(x^o(t), u^o(t), t) \right]^{\mathrm{T}} \lambda^o(t) .$

b) For all $t \in [t_a, t_b]$, the Hamiltonian $H(x^o(t), u, \lambda^o(t), \lambda_0^o, t)$ has a global minimum with respect to $u \in \Omega$ at $u = u^o(t)$, i.e.,

$\quad H(x^o(t), u^o(t), \lambda^o(t), \lambda_0^o, t) \leq H(x^o(t), u, \lambda^o(t), \lambda_0^o, t)$
$\quad$ for all $u \in \Omega$ and all $t \in [t_a, t_b]$.

c) Furthermore, if the final time t_b is free (Subproblem A.2):

$$H(x^o(t_b), u^o(t_b), \lambda^o(t_b), \lambda_0^o, t_b) = -\lambda_0^o \frac{\partial K}{\partial t}(t_b) .$$

2.1.3 Proof

According to the philosophy of the Lagrange multiplier method, the n-vector valued Lagrange multipliers λ_a, λ_b, and $\lambda(t)$, for $t = t_a, \ldots, t_b$, and the scalar Lagrange multiplier λ_0 are introduced. The latter either attains the value 1 in the regular case or the value 0 in the singular case. With these multipliers, the constraints of the optimal control problem can be adjoined to the original cost functional.

This leads to the following augmented cost functional:

$$\overline{J} = \lambda_0 K(t_b) + \int_{t_a}^{t_b} \left[\lambda_0 L(x(t), u(t), t) + \lambda(t)^{\mathrm{T}} \{ f(x(t), u(t), t) - \dot{x} \} \right] dt$$
$$+ \lambda_a^{\mathrm{T}} \{ x_a - x(t_a) \} + \lambda_b^{\mathrm{T}} \{ x_b - x(t_b) \} .$$

Introducing the Hamiltonian function

$$H(x(t), u(t), \lambda(t), \lambda_0, t) = \lambda_0 L(x(t), u(t), t) + \lambda(t)^\mathrm{T} f(x(t), u(t), t)$$

and dropping the notation of all of the independent variables allows us to write the augmented cost functional in the following rather compact form:

$$\overline{J} = \lambda_0 K(t_b) + \int_{t_a}^{t_b} \left[H - \lambda^\mathrm{T} \dot{x} \right] dt$$
$$+ \lambda_a^\mathrm{T} \{ x_a - x(t_a) \} + \lambda_b^\mathrm{T} \{ x_b - x(t_b) \} .$$

According to the philosophy of the Lagrange multiplier method, the augmented cost functional $\overline{J}$ has to be minimized with respect to all of its mutually independent variables $x(t_a)$, $x(t_b)$, λ_a, λ_b, and $u(t)$, $x(t)$, and $\lambda(t)$ for all $t \in (t_a, t_b)$, as well as t_b (if the final time is free). The two cases $\lambda_0 = 1$ and $\lambda_0 = 0$ have to be considered separately.

Suppose that we have found the optimal solution $x^o(t_a)$, $x^o(t_b)$, λ_a^o, λ_b^o, λ_0^o, and $u^o(t)$ (satisfying $u^o(t) \in \Omega$), $x^o(t)$, and $\lambda^o(t)$ for all $t \in (t_a, t_b)$, as well as t_b (if the final time is free).

The rules of differential calculus yield the following first differential $\delta \overline{J}$ of $\overline{J}(u^o)$ around the optimal solution:

$$\delta \overline{J} = \left[\lambda_0 \frac{\partial K}{\partial t} + H - \lambda^\mathrm{T} \dot{x} \right]_{t_b} \delta t_b$$
$$+ \int_{t_a}^{t_b} \left[\frac{\partial H}{\partial x} \delta x + \frac{\partial H}{\partial u} \delta u + \frac{\partial H}{\partial \lambda} \delta \lambda - \delta \lambda^\mathrm{T} \dot{x} - \lambda^\mathrm{T} \delta \dot{x} \right] dt$$
$$+ \delta \lambda_a^\mathrm{T} \{ x_a - x(t_a) \} - \lambda_a^\mathrm{T} \delta x(t_a)$$
$$+ \delta \lambda_b^\mathrm{T} \{ x_b - x(t_b) \} - \lambda_b^\mathrm{T} \left(\delta x + \dot{x} \delta t_b \right)_{t_b} .$$

Since we have postulated a minimum of the augmented function at $\overline{J}(u^o)$, this first differential must satisfy the inequality

$$\delta \overline{J} \geq 0$$

for all admissible variations of the independent variables. All of the variations of the independent variables are unconstrained, with the exceptions that $\delta u(t)$ is constrained to the tangent cone of Ω at $u^o(t)$, i.e.,

$$\delta u(t) \in T(\Omega, u^o(t)) \quad \text{for all } t \in [t_a, t_b] ,$$

such that the control constraint $u(t) \in \Omega$ is not violated, and

$$\delta t_b = 0$$

if the final time is fixed (Problem Type A.1).

However, it should be noted that $\delta\dot{x}(t)$ corresponds to $\delta x(t)$ differentiated with respect to time t. In order to remove this problem, the term $\int \lambda^{\mathrm{T}} \delta\dot{x}\, dt$ is integrated by parts. Thus, $\delta\dot{x}(t)$ will be replaced by $\delta x(t)$ and $\lambda(t)$ by $\dot{\lambda}(t)$. This yields

$$
\begin{aligned}
\delta\bar{J} = &\left[\lambda_0 \frac{\partial K}{\partial t} + H - \lambda^{\mathrm{T}}\dot{x} \right]_{t_b} \delta t_b - \left(\lambda^{\mathrm{T}}\delta x \right)_{t_b} + \left(\lambda^{\mathrm{T}}\delta x \right)_{t_a} \\
&+ \int_{t_a}^{t_b} \left[\frac{\partial H}{\partial x}\delta x + \frac{\partial H}{\partial u}\delta u + \frac{\partial H}{\partial \lambda}\delta\lambda - \delta\lambda^{\mathrm{T}}\dot{x} + \dot{\lambda}^{\mathrm{T}}\delta x \right] dt \\
&+ \delta\lambda_a^{\mathrm{T}}\{x_a - x(t_a)\} - \lambda_a^{\mathrm{T}}\delta x(t_a) \\
&+ \delta\lambda_b^{\mathrm{T}}\{x_b - x(t_b)\} - \lambda_b^{\mathrm{T}}\left(\delta x + \dot{x}\delta t_b \right)_{t_b} \\
= &\left[\lambda_0 \frac{\partial K}{\partial t} + H \right]_{t_b} \delta t_b \\
&+ \int_{t_a}^{t_b} \left[\left(\frac{\partial H}{\partial x} + \dot{\lambda}^{\mathrm{T}} \right) \delta x + \frac{\partial H}{\partial u}\delta u + \left(\frac{\partial H}{\partial \lambda} - \dot{x}^{\mathrm{T}} \right)\delta\lambda \right] dt \\
&+ \delta\lambda_a^{\mathrm{T}}\{x_a - x(t_a)\} + \left(\lambda^{\mathrm{T}}(t_a) - \lambda_a^{\mathrm{T}} \right)\delta x(t_a) \\
&+ \delta\lambda_b^{\mathrm{T}}\{x_b - x(t_b)\} - \left(\lambda^{\mathrm{T}}(t_b) + \lambda_b^{\mathrm{T}} \right)\left(\delta x + \dot{x}\delta t_b \right)_{t_b} \\
\geq\, &0 \quad \text{for all admissible variations.}
\end{aligned}
$$

According to the philosophy of the Lagrange multiplier method, this inequality must hold for arbitrary combinations of the mutually independent variations δt_b, and $\delta x(t)$, $\delta u(t)$, $\delta\lambda(t)$ at any time $t \in [t_a, t_b]$, and $\delta\lambda_a$, $\delta x(t_a)$, and $\delta\lambda_b$. Therefore, this inequality must be satisfied for a few very specially chosen combinations of these variations as well, namely where only one single variation is nontrivial and all of the others vanish.

The consequence is that all of the factors multiplying a differential must vanish.

There are two exceptions:

1) If the final time t_b is fixed, the final time must not be varied; therefore, the first bracketed term must only vanish if the final time is free.

2) If the optimal control $u^o(t)$ at time t lies in the interior of the control constraint set Ω, then the factor $\partial H/\partial u$ must vanish (and H must have a local minimum). If the optimal control $u^o(t)$ at time t lies on the boundary $\partial\Omega$ of Ω, then the inequality must hold for all $\delta u(t) \in T(\Omega, u^o(t))$. However, the gradient $\nabla_u H$ need not vanish. Rather, $-\nabla_u H$ is restricted to lie in the normal cone $T^*(\Omega, u^o(t))$, i.e., again, the Hamiltonian must have a (local) minimum at $u^o(t)$.

This completes the proof of Theorem A.

Notice that there are no conditions for λ_a and λ_b. In other words, the boundary conditions $\lambda^o(t_a)$ and $\lambda^o(t_b)$ of the optimal "costate" $\lambda^o(.)$ are free.

Remark: The calculus of variations only requests the local minimization of the Hamiltonian H with respect to the control u. — In Theorem A, the Hamiltonian is requested to be globally minimized over the admissible set Ω. This restriction is justified in Chapter 2.2.1.

2.1.4 Time-Optimal, Frictionless, Horizontal Motion of a Mass Point

Statement of the optimal control problem:

See Chapter 1.2, Problem 1, p. 5. — Since there is no friction and the final time t_b is not bounded, any arbitrary final state can be reached. There exists a unique optimal solution.

Using the cost functional $J(u) = \int_0^{t_b} dt$ leads to the Hamiltonian function

$$H = \lambda_0 + \lambda_1(t)x_2(t) + \lambda_2(t)u(t) \ .$$

Pontryagin's necessary conditions for optimality:

If $u^o : [0, t_b] \to [-a_{\max}, a_{\max}]$ is the optimal control and t_b the optimal final time, then there exists a nontrivial vector

$$\begin{bmatrix} \lambda_0^o \\ \lambda_1^o(t_b) \\ \lambda_2^o(t_b) \end{bmatrix} \neq \begin{bmatrix} 0 \\ 0 \\ 0 \end{bmatrix} \ ,$$

such that the following conditions are satisfied:

a) Differential equations and boundary conditions:

$$\dot{x}_1^o(t) = x_2^o(t)$$
$$\dot{x}_2^o(t) = u^o(t)$$
$$\dot{\lambda}_1^o(t) = -\frac{\partial H}{\partial x_1} = 0$$
$$\dot{\lambda}_2^o(t) = -\frac{\partial H}{\partial x_2} = -\lambda_1^o(t)$$
$$x_1^o(0) = s_a$$
$$x_2^o(0) = v_a$$
$$x_1^o(t_b) = s_b$$
$$x_2^o(t_b) = v_b \ .$$

b) Minimization of the Hamiltonian function:

$$H(x_1^o(t), x_2^o(t), u^o(t), \lambda_1^o(t), \lambda_2^o(t), \lambda_0^o) \leq H(x_1^o(t), x_2^o(t), u, \lambda_1^o(t), \lambda_2^o(t), \lambda_0^o)$$
$$\text{for all } u \in \Omega \text{ and all } t \in [0, t_b]$$

and hence

$$\lambda_2^o(t) u^o(t) \leq \lambda_2^o(t) u$$
$$\text{for all } u \in \Omega \text{ and all } t \in [0, t_b] \ .$$

c) At the optimal final time t_b:

$$H(t_b) = \lambda_0^o + \lambda_1^o(t_b) x_2^o(t_b) + \lambda_2^o(t_b) u^o(t_b) = 0 \ .$$

Minimizing the Hamiltonian function yields the following preliminary control law:

$$u^o(t) = \begin{cases} +a_{max} & \text{for } \lambda_2^o(t) < 0 \\ u \in \Omega & \text{for } \lambda_2^o(t) = 0 \\ -a_{max} & \text{for } \lambda_2^o(t) > 0 \ . \end{cases}$$

Note that for $\lambda_2^o(t) = 0$, every admissible control $u \in \Omega$ minimizes the Hamiltonian function.

Claim: The function $\lambda_2^o(t)$ has only isolated zeros, i.e., it cannot vanish on some interval $[a, b]$ with $b > a$.

Proof: The assumption $\lambda_2^o(t) \equiv 0$ leads to $\dot{\lambda}_2^o(t) \equiv 0$ and $\lambda_1^o(t) \equiv 0$. From the condition c at the final time t_b,

$$H(t_b) = \lambda_0^o + \lambda_1^o(t_b) x_2^o(t_b) + \lambda_2^o(t_b) u^o(t_b) = 0 \ ,$$

it follows that $\lambda_0^o = 0$ as well. — This contradiction with the nontriviality condition of Pontryagin's Minimum Principle proves the claim.

Therefore, we arrive at the following control law:

$$u^o(t) = -a_{max} \, \text{sign}\{\lambda_2^o(t)\} = \begin{cases} +a_{max} & \text{for } \lambda_2^o(t) < 0 \\ 0 & \text{for } \lambda_2^o(t) = 0 \\ -a_{max} & \text{for } \lambda_2^o(t) > 0 \ . \end{cases}$$

Of course, assigning the special value $u^o(t) = 0$ when $\lambda_2^o(t) = 0$ is arbitrary and has no special consequences.

Plugging this control law into the differential equation of x_2^o results in the two-point boundary value problem

$$\dot{x}_1^o(t) = x_2^o(t)$$
$$\dot{x}_2^o(t) = -a_{\max} \operatorname{sign}\{\lambda_2^o(t)\}$$
$$\dot{\lambda}_1^o(t) = 0$$
$$\dot{\lambda}_2^o(t) = -\lambda_1^o(t)$$
$$x_1^o(0) = s_a$$
$$x_2^o(0) = v_a$$
$$x_1^o(t_b) = s_b$$
$$x_2^o(t_b) = v_b \ ,$$

which needs to be solved. — Note that there are four differential equations with two boundary conditions at the initial time 0 and two boundary conditions at the (unknown) final time t_b.

The differential equations for the costate variables $\lambda_1^o(t)$ and $\lambda_2^o(t)$ imply that $\lambda_1^o(t) \equiv c_1^o$ is constant and that $\lambda_2^o(t)$ is an affine function of the time t:

$$\lambda_2^o(t) = -c_1^o t + c_2^o \ .$$

The remaining problem is finding the optimal values $(c_1^o, c_2^o) \neq (0, 0)$ such that the two-point boundary value problem is solved.

Obviously, the optimal open-loop control has the following features:

- Always, $|u^o(t)| \equiv a_{\max}$, i.e., there is always full acceleration or deceleration. This is called "bang-bang" control.
- The control switches at most once from $-a_{\max}$ to $+a_{\max}$ or from $+a_{\max}$ to $-a_{\max}$, respectively.

Knowing this simple structure of the optimal open-loop control, it is almost trivial to find the equivalent optimal closed-loop control with state feedback:

For a constant acceleration $u^o(t) \equiv a$ (where a is either $+a_{\max}$ or $-a_{\max}$), the corresponding state trajectory for $t > \tau$ is described in the parametrized form

$$x_2^o(t) = x_2^o(\tau) + a(t - \tau)$$
$$x_1^o(t) = x_1^o(\tau) + x_2^o(\tau)(t - \tau) + \frac{a}{2}(t - \tau)^2$$

or in the implicit form

$$x_1^o(t) - x_1^o(\tau) = \frac{x_2^o(\tau)}{a}\left(x_2^o(t) - x_2^o(\tau)\right) + \frac{1}{2a}\left(x_2^o(t) - x_2^o(\tau)\right)^2 \ .$$

In the state space (x_1, x_2) which is shown in Fig. 2.1, these equations define a segment on a parabola. The axis of the parabola coincides with the x_1 axis. For a positive acceleration, the parabola opens to the right and the state travels upward along the parabola. Conversely, for a negative acceleration, the parabola opens to the left and the state travels downward along the parabola.

The two parabolic arcs for $-a_{max}$ and $+a_{max}$ which end in the specified final state (s_b, v_b) divide the state space into two parts ("left" and "right").

The following optimal closed-loop state-feedback control law should now be obvious:

- $u^o(x_1, x_2) \equiv +a_{max}$ for all (x_1, x_2) in the open left part,
- $u^o(x_1, x_2) \equiv -a_{max}$ for all (x_1, x_2) in the open right part,
- $u^o(x_1, x_2) \equiv -a_{max}$ for all (x_1, x_2) on the left parabolic arc which ends at (s_b, v_b), and
- $u^o(x_1, x_2) \equiv +a_{max}$ for all (x_1, x_2) on the right parabolic arc which ends at (s_b, v_b).

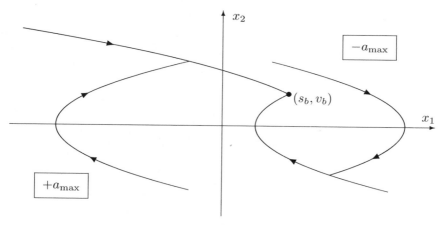

Fig. 2.1. Optimal feedback control law for the time-optimal motion.

2.1.5 Fuel-Optimal, Frictionless, Horizontal Motion of a Mass Point

Statement of the optimal control problem:

See Chapter 1.2, Problem 3, p. 6. — This problem has no solution, if the fixed final time t_b is too small, i.e., if the required transfer of the plant from the initial state (s_a, v_a) to the final state (s_b, v_b) is not possible in time. If the final time is sufficiently large, there exists at least one optimal solution.

Hamiltonian function:

$$H = |u(t)| + \lambda_1(t)x_2(t) + \lambda_2(t)u(t) \qquad \text{in the regular case with } \lambda_0^o = 1$$

$$H = \lambda_1(t)x_2(t) + \lambda_2(t)u(t) \qquad \text{in the singular case with } \lambda_0^o = 0.$$

Pontryagin's necessary conditions for optimality:

If $u^o : [0, t_b] \rightarrow [-a_{\max}, a_{\max}]$ is an optimal control, then there exists a nontrivial vector

$$\begin{bmatrix} \lambda_0^o \\ \lambda_1^o(t_b) \\ \lambda_2^o(t_b) \end{bmatrix} \neq \begin{bmatrix} 0 \\ 0 \\ 0 \end{bmatrix} ,$$

such that the following conditions are satisfied:

a) Differential equations and boundary conditions:

$$\dot{x}_1^o(t) = x_2^o(t)$$

$$\dot{x}_2^o(t) = u^o(t)$$

$$\dot{\lambda}_1^o(t) = -\frac{\partial H}{\partial x_1} = 0$$

$$\dot{\lambda}_2^o(t) = -\frac{\partial H}{\partial x_2} = -\lambda_1^o(t)$$

$$x_1^o(0) = s_a$$

$$x_2^o(0) = v_a$$

$$x_1^o(t_b) = s_b$$

$$x_2^o(t_b) = v_b .$$

b) Minimization of the Hamiltonian function:

$$H(x_1^o(t), x_2^o(t), u^o(t), \lambda_1^o(t), \lambda_2^o(t), \lambda_0^o) \leq H(x_1^o(t), x_2^o(t), u, \lambda_1^o(t), \lambda_2^o(t), \lambda_0^o)$$
$$\text{for all } u \in \Omega \text{ and all } t \in [0, t_b]$$

and hence

$$\lambda_0^o|u^o(t)| + \lambda_2^o(t)u^o(t) \leq \lambda_0^o|u| + \lambda_2^o(t)u$$
$$\text{for all } u \in \Omega \text{ and all } t \in [0, t_b] .$$

Minimizing the Hamiltonian function yields the following control law:

in the regular case with $\lambda_0^o = 1$:

$$u^o(t) = \begin{cases} +a_{\max} & \text{for } \lambda_2^o(t) < -1 \\ u \in [0, a_{\max}] & \text{for } \lambda_2^o(t) = -1 \\ 0 & \text{for } \lambda_2^o(t) \in (-1, +1) \\ u \in [-a_{\max}, 0] & \text{for } \lambda_2^o(t) = +1 \\ -a_{\max} & \text{for } \lambda_2^o(t) > +1 \end{cases}$$

and in the singular case with $\lambda_0^o = 0$:

$$u^o(t) = \begin{cases} +a_{\max} & \text{for } \lambda_2^o(t) < 0 \\ 0 & \text{for } \lambda_2^o(t) = 0 \\ -a_{\max} & \text{for } \lambda_2^o(t) > 0 \ . \end{cases}$$

From the analysis of Problem 1 in Chapter 2.1.4, we already know that $\lambda_2^o(t)$ can only have a zero at a discrete time in the singular case. Therefore, assigning $u^o(t){=}0$ for $\lambda_2^o(t){=}0$ is acceptable.

Again, the differential equations for the costate variables $\lambda_1^o(t)$ and $\lambda_2^o(t)$ imply that $\lambda_1^o(t) \equiv c_1^o$ is constant and that $\lambda_2^o(t)$ is an affine function of the time t:

$$\lambda_2^o(t) = -c_1^o t + c_2^o \ .$$

Clearly, for $c_1^o \neq 0$, the costate variable $\lambda_2^o(t)$ can only attain the values $+1$ and -1 at one discrete time, each. In this, case we may arbitrarily assign $u^o(t){=}0$ for $\lambda_2^o(t){=}\pm 1$. — However, there are constellations concerning the initial state (s_a, v_a) and the final state (s_b, v_b), where $\lambda_2^o(t){\equiv}-1$ or $\lambda_2^o(t){\equiv}+1$ is necessary.

Remarks:

- The singular case with $\lambda_0^o{=}0$ is only necessary, if the final time t_b corresponds to the final time $t_{b,\min}$ of the time-optimal case and if the optimal control needs to switch. In this case, the optimal control law is:

$$u^o(\lambda_2^o(t)) = -a_{\max} \operatorname{sign}\{\lambda_2^o(t)\} \ .$$

- In all other cases, we have $\lambda_0^0 = 1$.

- If $t_b > t_{b,\min}$, there are cases where $\lambda_2^o(t) \equiv -1$ or $\lambda_2^o(t) \equiv +1$ is necessary. Then, there are infinitely many optimal solutions. They are characterized by the fact that only acceleration or deceleration occurs in order to arrive at the specified terminal state in time.

- In the remaining (more general) cases with $t_b > t_{b,\min}$, we obtain the following optimal control law with $c_1^o \neq 0$:

$$
u^o(\lambda_2^o(t)) = \begin{cases} +a_{\max} & \text{for } \lambda_2^o(t) < -1 \\ 0 & \text{for } \lambda_2^o(t) \in [-1, 1] \\ -a_{\max} & \text{for } \lambda_2^o(t) > +1 \ . \end{cases}
$$

The resulting two-point boundary value problem

$$
\begin{aligned}
\dot{x}_1^o(t) &= x_2^o(t) \\
\dot{x}_2^o(t) &= u^o\{\lambda_2^o(t)\} \\
\lambda_1^o(t) &\equiv c_1^o \\
\lambda_2^o(t) &= c_2^o - c_1^o t \\
x_1^o(0) &= s_a \\
x_2^o(0) &= v_a \\
x_1^o(t_b) &= s_b \\
x_2^o(t_b) &= v_b
\end{aligned}
$$

remains to be solved by finding the optimal values c_1^o and c_2^o.

In the general case, where the optimal solution is unique, it has the following features:

- Always, $|u^o(t)| \equiv a_{\max}$ or 0, i.e., there is always full acceleration, or full deceleration, or the speed x_2 is constant. This may be call "boost-sustain-boost" control.
- The control switches at most once each from $-a_{\max}$ to 0 and from 0 to $+a_{\max}$, or from $+a_{\max}$ to 0 and from 0 to $-a_{\max}$, respectively.
- In the state space (x_1, x_2), the optimal trajectory consists at most of one parabolic arc with $u^o \equiv \pm a_{\max}$, and one parabolic arc with $u^o \equiv \mp a_{\max}$, and a straight line parallel to the x_1 axis in-between (see Fig. 2.2).
- In the singular case, the fuel-optimal trajectory degenerates to the time-optimal one. The middle straight-line section with $u^o \equiv 0$ vanishes.

The construction of the fuel-optimal state trajectory proceeds as follows: First, choose the pair of parabolic arcs starting at the initial state (s_a, v_a) and ending at the final state (s_b, v_b), respectively, which corresponds to the time-optimal motion (see Fig. 2.1). Then, choose the straight-line section between these two parabolic arcs, such that the moving state (x_1, x_2) arrives at the final state (s_b, v_b) just in time.

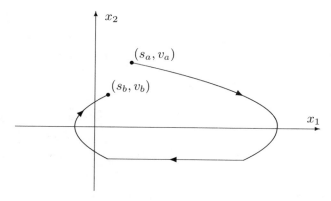

Fig. 2.2. Optimal feedback control law for the fuel-optimal motion.

2.2 Some Fine Points

2.2.1 Strong Control Variation and Global Minimization of the Hamiltonian

In Chapter 2.1.3, we have derived the necessary condition that the Hamiltonian function H must have a local minimum with respect to u at $u^o(t)$ for all $t \in [t_a, t_b]$ because we have considered a "weak" variation $\delta u(\cdot)$ which is small in the function space $L_m^\infty(t_a, t_b)$ and which contributes a small variation

$$\int_{t_a}^{t_b} \frac{\partial H}{\partial u} \delta u \, dt$$

to the differential $\delta \overline{J}$.

Alternatively, we may consider a "strong" variation $\delta u[u, \tau, \delta\tau](\cdot)$ of the form

$$\delta u[u, \tau, \delta\tau](t) = \begin{cases} 0 & \text{for } t < \tau, \ \tau \in (t_a, t_b) \\ u - u^o(t) & \text{for } \tau \leq t < \tau + \delta\tau, \ u \in \Omega \\ 0 & \text{for } \tau + \delta\tau \leq t \leq t_b \end{cases}$$

which is small in the function space $L_m^1(t_a, t_b)$. Its contribution to the differential $\delta \overline{J} \geq 0$ is:

$$\Big(H(x^o(\tau), u, \lambda^o(\tau), \lambda_0^o, \tau) - H(x^o(\tau), u^o(\tau), \lambda^o(\tau), \lambda_0^o, \tau) \Big) \delta\tau \ .$$

Since this contribution must be non-negative for all $u \in \Omega$ and for all $\tau \in [t_a, t_b)$ and since $\delta\tau \geq 0$, it follows that the Hamiltonian function must be globally minimized, as formulated in Theorem A.

2.2.2 Evolution of the Hamiltonian

Consider the value of the Hamiltonian function along an optimal trajectory according to the time-varying function $H^o : [t_a, t_b] \to R$:

$$H^o(t) = H(x^o(t), u^o(t), \lambda^o(t), \lambda^o_0, t) \ .$$

Its total derivative with respect to the time t is:

$$\frac{dH^o(t)}{dt} = \frac{\partial H|_o}{\partial x} \dot{x}^o(t) + \frac{\partial H|_o}{\partial \lambda} \dot{\lambda}^o(t) + \frac{\partial H|_o}{\partial u} \dot{u}^o(t) + \frac{\partial H|_o}{\partial t} \ .$$

According to Pontryagin's Minimum Principle, the first two terms cancel. The third term vanishes due to the minimization of the Hamiltonian function. This is true, even if $u^o(t)$ lies on the boundary $\partial\Omega$ of the control constraint set Ω, because the latter is time-invariant and because the gradient $\nabla_u H|_o$ is transversal to $\partial\Omega$, i.e., $-\nabla_u H|_o$ lies in the normal cone $T^*(\Omega, u^o(t))$ of the tangent cone $T(\Omega, u^o(t))$ of Ω at $u^o(t)$.

Therefore, the total derivative of the Hamiltonian function along an optimal trajectory is identical to its partial derivative:

$$\frac{dH^o(t)}{dt} = \frac{\partial H}{\partial t}(x^o(t), u^o(t), \lambda^o(t), \lambda^o_0, t) \ .$$

In the special case of a time-invariant optimal control problem, this leads to:

$$H \equiv \text{constant} \qquad \text{if the final time } t_b \text{ is fixed,}$$

$$H \equiv 0 \qquad \text{if the final time } t_b \text{ is free.}$$

These facts are trivial. But sometimes, they are useful in the analysis of singular optimal control problems (see Chapter 2.6) or for the investigation of jump discontinuities of the optimal solution (see Chapter 2.5.4).

2.2.3 Special Case: Cost Functional $J(u) = \pm x_i(t_b)$

In the formulation of the most general optimal control problem in Chapter 1.1, it has been implicitly assumed that the integrand $L(x, u, t)$ of the cost functional is not identical to any of the functions $f_i(x, u, t)$ of the vector function $f(x, u, t)$ in the state differential equation.

However, if the cost functional is of the form

$$J(u) = x_i(t_b) = \int_{t_a}^{t_b} f_i(x(t), u(t), t) \, dt \ ,$$

this assumption is violated. In this case, the standard problem statement is devious in the sense that the nontriviality condition

$$\begin{bmatrix} \lambda_0^o \\ \lambda^o(t_b) \end{bmatrix} \neq 0 \in R^{n+1}$$

of Pontryagin's Minimum Principle is useless: The forbidden situation

$$\lambda_0 = \lambda_1(t_b) = \ldots = \lambda_i(t_b) = \ldots = \lambda_n(t_b) = 0$$

is equivalent to the seemingly acceptable situation

$$\lambda_0 = 1$$
$$\lambda_i(t_b) = -1$$
$$\lambda_j(t_b) = 0 \text{ for all } j \neq i .$$

Problem solving technique for $J(u) = x_i(t_b)$:

The additional Lagrange multiplier λ_0 for the cost functional is no longer needed. The correct Hamiltonian function in this case is: $H : R^n \times \Omega \times R^n \times [t_a, t_b] \to R^n$,

$$H(x(t), u(t), \lambda(t), t) = \lambda^T(t) f(x(t), u(t), t) .$$

The obvious new nontriviality condition is:

$$\lambda^o(t_b) \neq 0 \in R^n \text{ with } \lambda_i^o(t_b) = 1 \text{ or } 0 .$$

Problem solving technique for $J(u) = -x_i(t_b)$:

The minimization of $-x_i(t_b)$ is equivalent to the maximization of $x_i(t_b)$. Therefore, there are two routes: Either we minimize $J(u) = -x_i(t_b)$ and use Pontryagin's Minimum Principle, or we maximize $J(u) = x_i(t_b)$ and use Pontryagin's Maximum Principle, i.e., we globally maximize the Hamiltonian function rather than minimizing it.

For more details and for a highly rigorous treatment of optimal control problems with a cost functional of the form $J(u) = x_1(t_b)$, the reader is referred to [20]. — See also Chapter 1.2, Problem 4 and Chapter 2.8.4.

2.3 Optimal Control Problems with a Free Final State

In this section, Pontryagin's Minimum Principle is derived for optimal control problems with a completely unspecified final state (and no state constraints). For illustration, the LQ regulator problem is solved.

2.3.1 The Optimal Control Problem of Type C

Statement of the optimal control problem:

Find a piecewise continuous control $u : [t_a, t_b] \rightarrow \Omega \subseteq R^m$, such that the constraints

$$x(t_a) = x_a$$

$$\dot{x}(t) = f(x(t), u(t), t) \qquad \text{for all } t \in [t_a, t_b]$$

are satisfied and such that the cost functional

$$J(u) = K(x(t_b), t_b) + \int_{t_a}^{t_b} L(x(t), u(t), t)\, dt$$

is minimized;

Subproblem C.1: t_b is fixed,

Subproblem C.2: t_b is free $(t_b > t_a)$.

Remark: t_a, $x_a \in R^n$ are specified; $\Omega \subseteq R^m$ is time-invariant.

2.3.2 Pontryagin's Minimum Principle

Definition: Hamiltonian function $H : R^n \times \Omega \times R^n \times [t_a, t_b] \rightarrow R$,

$$H(x(t), u(t), \lambda(t), t) = L(x(t), u(t), t) + \lambda^{\mathrm{T}}(t) f(x(t), u(t), t) .$$

Note: For the problem of Type C, the singular case with $\lambda_0^o = 0$ cannot occur.

Theorem C

If the control $u^o : [t_a, t_b] \rightarrow \Omega$ is optimal, then the following conditions are satisfied:

a) $\dot{x}^o(t) = \nabla_\lambda H_{|o} = f(x^o(t), u^o(t), t)$

  $x^o(t_a) = x_a$

  $\dot{\lambda}^o(t) = -\nabla_x H_{|o} = -\nabla_x L(x^o(t), u^o(t), t) - \left[\dfrac{\partial f}{\partial x}(x^o(t), u^o(t), t) \right]^{\mathrm{T}} \lambda^o(t)$

  $\lambda^o(t_b) = \nabla_x K(x^o(t_b), t_b) .$

b) For all $t \in [t_a, t_b]$ the Hamiltonian $H(x^o(t), u, \lambda^o(t), t)$ has a global minimum with respect to $u \in \Omega$ at $u = u^o(t)$, i.e.,

$$H(x^o(t), u^o(t), \lambda^o(t), t) \leq H(x^o(t), u, \lambda^o(t), t)$$
for all $u \in \Omega$ and all $t \in [t_a, t_b]$.

c) Furthermore, if the final time t_b is free (Subproblem C.2):

$$H(x^o(t_b), u^o(t_b), \lambda^o(t_b), t_b) = -\frac{\partial K}{\partial t}(x^o(t_b), t_b) .$$

2.3.3 Proof

Proving Theorem C proceeds in complete analogy to the proof of Theorem A given in Chapter 2.1.3.

Since singularity with $\lambda_0^o = 0$ cannot occur in optimal control problems of Type C, the augmented cost functional is:

$$\overline{J} = K(x(t_b), t_b) + \int_{t_a}^{t_b} \left[L(x, u, t) + \lambda(t)^{\mathrm{T}} \{ f(x, u, t) - \dot{x} \} \right] dt + \lambda_a^{\mathrm{T}} \{ x_a - x(t_a) \}$$

$$= K(x(t_b), t_b) + \int_{t_a}^{t_b} \left[H - \lambda^{\mathrm{T}} \dot{x} \right] dt + \lambda_a^{\mathrm{T}} \{ x_a - x(t_a) \} ,$$

where $H = H(x, u, \lambda, t) = L(x, u, t) + \lambda^{\mathrm{T}} f(x, u, t)$ is the Hamiltonian function.

According to the philosophy of the Lagrange multiplier method, the augmented cost functional $\overline{J}$ has to be minimized with respect to all of its mutually independent variables $x(t_a)$, λ_a, $x(t_b)$, and $u(t)$, $x(t)$, and $\lambda(t)$ for all $t \in (t_a, t_b)$, as well as t_b (if the final time is free).

Suppose, we have found the optimal solution $x^o(t_a)$, λ_a^o, $x^o(t_b)$, and $u^o(t)$ (satisfying $u^o(t) \in \Omega$), $x^o(t)$, and $\lambda^o(t)$ for all $t \in (t_a, t_b)$, as well as t_b (if the final time is free).

The following first differential $\delta\overline{J}$ of $\overline{J}(u^o)$ around the optimal solution is obtained (for details of the analysis, consult Chapter 2.1.3):

$$\delta\overline{J} = \left[\left(\frac{\partial K}{\partial x} - \lambda^{\mathrm{T}} \right) (\delta x + \dot{x} \delta t_b) \right]_{t_b} + \left[\frac{\partial K}{\partial t} + H \right]_{t_b} \delta t_b$$

$$+ \int_{t_a}^{t_b} \left[\left(\frac{\partial H}{\partial x} + \dot{\lambda}^{\mathrm{T}} \right) \delta x + \frac{\partial H}{\partial u} \delta u + \left(\frac{\partial H}{\partial \lambda} - \dot{x}^{\mathrm{T}} \right) \delta\lambda \right] dt$$

$$+ \delta\lambda_a^{\mathrm{T}} \{ x_a - x(t_a) \} + \left(\lambda^{\mathrm{T}}(t_a) - \lambda_a^{\mathrm{T}} \right) \delta x(t_a) .$$

Since we have postulated a minimum of the augmented function at $\overline{J}(u^o)$, this first differential must satisfy the inequality

$$\delta \overline{J} \geq 0$$

for all admissible variations of the independent variables. All of the variations of the independent variables are unconstrained, with the exceptions that $\delta u(t)$ is constrained to the tangent cone of Ω at $u^o(t)$, i.e.,

$$\delta u(t) \in T(\Omega, u^o(t)) \quad \text{for all } t \in [t_a, t_b] \ ,$$

such that the control constraint $u(t) \in \Omega$ is not violated, and

$$\delta t_b = 0$$

if the final time is fixed (Problem Type C.1).

According to the philosophy of the Lagrange multiplier method, this inequality must hold for arbitrary combinations of the mutually independent variations δt_b, and $\delta x(t)$, $\delta u(t)$, $\delta \lambda(t)$ at any time $t \in (t_a, t_b)$, and $\delta \lambda_a$, $\delta x(t_a)$, and $\delta x(t_b)$. Therefore, this inequality must be satisfied for a few very specially chosen combinations of these variations as well, namely where only one single variation is nontrivial and all of the others vanish.

The consequence is that all of the factors multiplying a differential must vanish.

There are two exceptions:

1) If the final time is fixed, it must not be varied; therefore, the second bracketed term must only vanish if the final time is free.

2) If the optimal control $u^o(t)$ at time t lies in the interior of the control constraint set Ω, then the factor $\partial H/\partial u$ must vanish (and H must have a local minimum). If the optimal control $u^o(t)$ at time t lies on the boundary $\partial \Omega$ of Ω, then the inequality must hold for all $\delta u(t) \in T(\Omega, u^o(t))$. However, the gradient $\nabla_u H$ need not vanish. Rather, $-\nabla_u H$ is restricted to lie in the normal cone $T^*(\Omega, u^o(t))$, i.e., again, the Hamiltonian must have a (local) minimum at $u^o(t)$.

This completes the proof of Theorem C.

Notice that there is no condition for λ_a. In other words, the boundary condition $\lambda^o(t_a)$ of the optimal costate $\lambda^o(.)$ is free.

Remark: The calculus of variations only requests the local minimization of the Hamiltonian H with respect to the control u. — In Theorem C, the Hamiltonian is requested to be globally minimized over the admissible set Ω. This restriction is justified in Chapter 2.2.1.

2.3.4 The LQ Regulator Problem

Statement of the optimal control problem:

See Chapter 1.2, Problem 5, p. 8. — This problem has a unique optimal solution (see Chapter 2.7, Theorem 1).

Hamiltonian function:

$$H = \frac{1}{2}x^{\mathrm{T}}(t)Q(t)x(t) + \frac{1}{2}u^{\mathrm{T}}(t)R(t)u(t) + \lambda^{\mathrm{T}}(t)A(t)x(t) + \lambda^{\mathrm{T}}(t)B(t)u(t) \ .$$

Pontryagin's necessary conditions for optimality:

If $u^o : [t_a, t_b] \to R^m$ is the optimal control, then the following conditions are satisfied:

a) Differential equations and boundary conditions:

$$\dot{x}^o(t) = \nabla_\lambda H = A(t)x^o(t) + B(t)u^o(t)$$
$$\dot{\lambda}^o(t) = -\nabla_x H = - Q(t)x^o(t) - A^{\mathrm{T}}(t)\lambda^o(t)$$
$$x^o(t_a) = x_a$$
$$\lambda^o(t_b) = \nabla_x K = Fx^o(t_b) \ .$$

b) Minimization of the Hamiltonian function:

$$H(x^o(t), u^o(t), \lambda^o(t), t) \leq H(x^o(t), u, \lambda^o(t), t)$$
$$\text{for all } u \in R^m \text{ and all } t \in [t_a, t_b]$$

and hence

$$\nabla_u H = R(t)u^o(t) + B^{\mathrm{T}}(t)\lambda^o(t) = 0 \ .$$

Minimizing the Hamiltonian function yields the following optimal open-loop control law:
$$u^o(t) = -R^{-1}(t)B^{\mathrm{T}}(t)\lambda^o(t) \ .$$

Plugging this control law into the differential equation of x results in the linear two-point boundary value problem

$$\dot{x}^o(t) = A(t)x^o(t) - B(t)R^{-1}(t)B^{\mathrm{T}}\lambda^o(t)$$
$$\dot{\lambda}^o(t) = - Q(t)x^o(t) - A^{\mathrm{T}}(t)\lambda^o(t)$$
$$x^o(t_a) = x_a$$
$$\lambda^o(t_b) = Fx^o(t_b)$$

which needs to be solved.

The two differential equations are homogeneous in $(x^o; \lambda^o)$ and at the final time t_b, the costate vector $\lambda(t_b)$ is a linear function of the final state vector $x^o(t_b)$. This leads to the conjecture that the costate vector might be a linear function of the state vector at all times.

Therefore, we try the linear ansatz

$$\lambda^o(t) = K(t)x^o(t) \; ,$$

where $K(t)$ is a suitable time-varying n by n matrix.

Differentiating this equation with respect to the time t, and considering the differential equations for the costate λ and the state x, and applying the ansatz in the differential equations leads to the following equation:

$$\dot{\lambda} = \dot{K}x + K\dot{x} = \dot{K}x + K(A - BR^{-1}B^{\mathrm{T}}K)x = -Qx - A^{\mathrm{T}}Kx$$

or equivalently to:

$$\left(\dot{K} + A^{\mathrm{T}}K + KA - KBR^{-1}B^{\mathrm{T}}K + Q \right) x \equiv 0 \; .$$

This equation must be satisfied at all times $t \in [t_a, t_b]$. Furthermore, we arrive at this equation, irrespective of the initial state x_a at hand, i.e., for all $x_a \in R^n$. Thus, the vector x in this equation may be an arbitrary vector in R^n. Therefore, the sum of matrices in the brackets must vanish.

The result is the optimal state feedback control law

$$u^o(t) = -G(t)x^o(t) = -R^{-1}(t)B^{\mathrm{T}}(t)K(t)x^o(t) \; ,$$

where the symmetric and positive-(semi)definite n by n matrix $K(t)$ is the solution of the matrix Riccati differential equation

$$\dot{K}(t) = -A^{\mathrm{T}}(t)K(t) - K(t)A(t) + K(t)B(t)R^{-1}(t)B^{\mathrm{T}}(t)K(t) - Q(t)$$

with the boundary condition

$$K(t_b) = F$$

at the final time t_b. Thus, the optimal time-varying gain matrix $G(t) = R^{-1}B^{\mathrm{T}}(t)K(t)$ of the state feedback controller can (and must) be computed and stored in advance.

In all of the textbooks about optimal control of linear systems, the LQ problem presented here is extended or specialized to the case of a time-invariant system with constant matrices A and B, and to constant penalty matrices Q and R, and to "infinite horizon", i.e., for the time interval $[t_a, t_b] = [0, \infty]$. This topic is not pursued here. The reader is referred to standard textbooks on linear optimal control, such as [1], [11], [16], and [25].

2.4 Optimal Control Problems with a Partially Constrained Final State

In this section, Pontryagin's Minimum Principle is derived for optimal control problems with a partially specified final state (and no state constraints). In other words, the final state $x^o(t_b)$ is restricted to lie in a closed "target set" $S \subseteq R^n$. — Obviously, the Problem Types A and C are special cases of the Type B with $S = \{x_b\}$ and $S = R^n$, respectively.

2.4.1 The Optimal Control Problem of Type B

Statement of the optimal control problem:

Find a piecewise continuous control $u : [t_a, t_b] \rightarrow \Omega \subseteq R^m$, such that the constraints

$$x(t_a) = x_a$$
$$\dot{x}(t) = f(x(t), u(t), t) \qquad \text{for all } t \in [t_a, t_b]$$
$$x(t_b) \in S$$

are satisfied and such that the cost functional

$$J(u) = K(x(t_b), t_b) + \int_{t_a}^{t_b} L(x(t), u(t), t)\, dt$$

is minimized;

Subproblem B.1: t_b is fixed,

Subproblem B.2: t_b is free $(t_b > t_a)$.

Remark: t_a, $x_a \in R^n$, $S \subseteq R^n$ are specified; $\Omega \subseteq R^m$ is time-invariant.

2.4.2 Pontryagin's Minimum Principle

Definition: Hamiltonian function $H : R^n \times \Omega \times R^n \times \{0, 1\} \times [t_a, t_b] \rightarrow R$,

$$H(x(t), u(t), \lambda(t), \lambda_0, t) = \lambda_0 L(x(t), u(t), t) + \lambda^{\mathrm{T}}(t) f(x(t), u(t), t) \ .$$

Theorem B

If the control $u^o : [t_a, t_b] \rightarrow \Omega$ is optimal, then there exists a nontrivial vector

$$\begin{bmatrix} \lambda_0^o \\ \lambda^o(t_b) \end{bmatrix} \neq 0 \in R^{n+1} \quad \text{with } \lambda_0^o = \begin{cases} 1 & \text{in the regular case} \\ 0 & \text{in the singular case}, \end{cases}$$

such that the following conditions are satisfied:

a) $\dot{x}^o(t) = \nabla_\lambda H_{|o} = f(x^o(t), u^o(t), t)$

   $x^o(t_a) = x_a$

   $$\dot{\lambda}^o(t) = -\nabla_x H_{|o} = -\lambda_0^o \nabla_x L(x^o(t), u^o(t), t) - \left[\frac{\partial f}{\partial x}(x^o(t), u^o(t), t)\right]^{\mathrm{T}} \lambda^o(t)$$

   $\lambda^o(t_b) = \lambda_0^o \nabla_x K(x^o(t_b), t_b) + q^o \qquad$ with $\qquad q^o \in T^*(S, x^o(t_b))^1$.

b) For all $t \in [t_a, t_b]$, the Hamiltonian $H(x^o(t), u, \lambda^o(t), \lambda_0^o, t)$ has a global minimum with respect to $u \in \Omega$ at $u = u^o(t)$, i.e.,

   $H(x^o(t), u^o(t), \lambda^o(t), \lambda_0^o, t) \leq H(x^o(t), u, \lambda^o(t), \lambda_0^o, t)$
   for all $u \in \Omega$ and all $t \in [t_a, t_b]$.

c) Furthermore, if the final time t_b is free (Subproblem B.2):

   $$H(x^o(t_b), u^o(t_b), \lambda^o(t_b), \lambda_0^o, t_b) = -\lambda_0^o \frac{\partial K}{\partial t}(x^o(t_b), t_b) .$$

2.4.3 Proof

Proving Theorem B proceeds in complete analogy to the proof of Theorem A given in Chapter 2.1.3.

With $\lambda_0^o = 1$ in the regular case and $\lambda_0^o = 0$ in the singular case, the augmented cost functional is:

$$\overline{J} = \lambda_0 K(x(t_b), t_b)$$
$$+ \int_{t_a}^{t_b} \left[\lambda_0 L(x, u, t) + \lambda(t)^{\mathrm{T}}\{f(x, u, t) - \dot{x}\}\right] dt + \lambda_a^{\mathrm{T}}\{x_a - x(t_a)\}$$
$$= \lambda_0 K(x(t_b), t_b) + \int_{t_a}^{t_b} \left[H - \lambda^{\mathrm{T}}\dot{x}\right] dt + \lambda_a^{\mathrm{T}}\{x_a - x(t_a)\} ,$$

where $H = H(x, u, \lambda, \lambda_0, t) = \lambda_0 L(x, u, t) + \lambda^{\mathrm{T}} f(x, u, t)$ is the Hamiltonian function.

According to the philosophy of the Lagrange multiplier method, the augmented cost functional $\overline{J}$ has to be minimized with respect to all of its mutually independent variables $x(t_a)$, λ_a, $x(t_b)$, and $u(t)$, $x(t)$, and $\lambda(t)$ for all $t \in (t_a, t_b)$, as well as t_b (if the final time is free).

Suppose, we have found the optimal solution $x^o(t_a)$, λ_a^o, $x^o(t_b)$ (satisfying $x^o(t_b) \in S$), and $u^o(t)$ (satisfying $u^o(t) \in \Omega$), $x^o(t)$, and $\lambda^o(t)$ for all $t \in (t_a, t_b)$, as well as t_b (if the final time is free).

---

[1] Normal cone of the tangent cone $T(S, x^o(t_b))$ of S at $x^o(t_b)$. This is the so-called transversality condition.

The following first differential $\delta\overline{J}$ of $\overline{J}(u^o)$ around the optimal solution is obtained (for details of the analysis, consult Chapter 2.1.3):

$$
\delta\overline{J} = \left[\left(\lambda_0\frac{\partial K}{\partial x} - \lambda^{\mathrm{T}}\right)(\delta x + \dot{x}\delta t_b)\right]_{t_b} + \left[\lambda_0\frac{\partial K}{\partial t} + H\right]_{t_b}\delta t_b
$$
$$
+ \int_{t_a}^{t_b}\left[\left(\frac{\partial H}{\partial x} + \dot{\lambda}^{\mathrm{T}}\right)\delta x + \frac{\partial H}{\partial u}\delta u + \left(\frac{\partial H}{\partial \lambda} - \dot{x}^{\mathrm{T}}\right)\delta\lambda\right]dt
$$
$$
+ \delta\lambda_a^{\mathrm{T}}\{x_a - x(t_a)\} + \left(\lambda^{\mathrm{T}}(t_a) - \lambda_a^{\mathrm{T}}\right)\delta x(t_a) .
$$

Since we have postulated a minimum of the augmented function at $\overline{J}(u^o)$, this first differential must satisfy the inequality

$$
\delta\overline{J} \geq 0
$$

for all admissible variations of the independent variables. All of the variations of the independent variables are unconstrained, with the exceptions that $\delta u(t)$ is constrained to the tangent cone of Ω at $u^o(t)$, i.e.,

$$
\delta u(t) \in T(\Omega, u^o(t)) \quad \text{for all } t \in [t_a, t_b] ,
$$

such that the control constraint $u(t) \in \Omega$ is not violated, and that $\delta x(t_b)$ is constrained to the tangent cone of S at $x^o(t_b)$, i.e.,

$$
\delta x(t_b) \in T(S, x^o(t_b)) ,
$$

such that the constraint $x(t_b) \in S$ is not violated, and

$$
\delta t_b = 0
$$

if the final time is fixed (Problem Type B.1).

According to the philosophy of the Lagrange multiplier method, this inequality must hold for arbitrary combinations of the mutually independent variations δt_b, and $\delta x(t)$, $\delta u(t)$, $\delta\lambda(t)$ at any time $t \in (t_a, t_b)$, and $\delta\lambda_a$, $\delta x(t_a)$, and $\delta x(t_b)$. Therefore, this inequality must be satisfied for a few very specially chosen combinations of these variations as well, namely where only one single variation is nontrivial and all of the others vanish.

The consequence is that all of the factors multiplying a differential must vanish.

There are three exceptions:

1) If the final time is fixed, it must not be varied; therefore, the second bracketed term must only vanish if the final time is free.

2) If the optimal control $u^o(t)$ at time t lies in the interior of the control constraint set Ω, then the factor $\partial H/\partial u$ must vanish (and H must have a local minimum). If the optimal control $u^o(t)$ at time t lies on the boundary $\partial\Omega$ of Ω, then the inequality must hold for all $\delta u(t) \in T(\Omega, u^o(t))$. However, the gradient $\nabla_u H$ need not vanish. Rather, $-\nabla_u H$ is restricted to lie in the normal cone $T^*(\Omega, u^o(t))$, i.e., again, the Hamiltonian must have a (local) minimum at $u^o(t)$.

3) If the optimal final state $x^o(t_b)$ lies in the interior of the target set S, then the factor in the first round brackets must vanish. If the optimal final state $x^o(t_b)$ lies on the boundary ∂S of S, then the inequality must hold for all $\delta x(t_b) \in T(S, x^o(t_b))$. In other words, $\lambda^o(t_b)$ can be of the form

$$\lambda^o(t_b) = \lambda_0^o \nabla_x K(x^o(t_b), t_b) + q \,,$$

where q must lie in the normal cone $T^*(S, x^o(t_b))$ of the target set S at $x^o(t_b)$. This guarantees that the resulting term satisfies

$$-q^{\mathrm{T}} \delta x(t_b) \geq 0$$

for all permissible variations $\delta x(t_b)$ of the final state $x^o(t_b)$.

This completes the proof of Theorem B.

Notice that there is no condition for λ_a. In other words, the boundary condition $\lambda^o(t_a)$ of the optimal costate $\lambda^o(.)$ is free.

Remark: The calculus of variations only requests the local minimization of the Hamiltonian H with respect to the control u. — In Theorem B, the Hamiltonian is requested to be globally minimized over the admissible set Ω. This restriction is justified in Chapter 2.2.1.

2.4.4 Energy-Optimal Control

Statement of the optimal control problem:

Consider the following energy-optimal control problem for an unstable system: Find $u : [0, t_b] \to R$, such that the system

$$\dot{x}(t) = ax(t) + bu(t) \quad \text{with } a > 0 \text{ and } b > 0$$

is transferred from the initial state

$$x(0) = x_0 > 0$$

to the final state

$$0 \leq x(t_b) \leq c\,, \quad \text{where } c < e^{at_b}x_0\,,$$

at the fixed final time t_b and such that the cost functional

$$J(u) = \int_0^{t_b} \frac{1}{2} u^2(t) \, dt$$

is minimized.

Since the partially specified final state $x^o(t_b)$ lies within the set of all of the reachable states at the final time t_b, a non-singular optimal solution exists.

Hamiltonian function:

$$H = \frac{1}{2} u^2(t) + \lambda(t) a x(t) + \lambda(t) b u(t) \ .$$

Pontryagin's necessary conditions for optimality:

If $u^o : [0, t_b] \to R$ is an optimal control, then the following conditions are satisfied:

a) Differential equations and boundary conditions:

$$\dot{x}^o(t) = \nabla_\lambda H = a x^o(t) + b u^o(t)$$

$$\dot{\lambda}^o(t) = -\nabla_x H = -a \lambda^o(t)$$

$$x^o(0) = x_0$$

$$x^o(t_b) \in S$$

$$\lambda^o(t_b) = q^o \in T^*(S, x^o(t_b)) \ .$$

b) Minimization of the Hamiltonian function:

$$u^o(t) + b \lambda^o(t) = 0 \quad \text{for all } t \in [0, t_b] \ .$$

Since the system is unstable and $c < e^{a t_b} x_0$, it is clear that the optimal final state lies at the upper boundary c of the specified target set $S = [0, c]$.

According to Pontryagin's Minimum Principle, the costate trajectory $\lambda(.)$ is described by

$$\lambda^o(t) = e^{-at} \lambda^o(0) \ ,$$

where $\lambda^o(0)$ is its unknown initial condition. Therefore, at the final time t_b, we have:

$$\lambda^o(t_b) = q^0 = e^{-a t_b} \lambda^o(0) > 0$$

as required, provided $\lambda^o(0) > 0$.

Using the optimal open-loop control law

$$u^o(t) = -b \lambda^o(t) = -b e^{-at} \lambda^o(0) \ ,$$

the unknown initial condition $\lambda^o(0)$ can be determined from the boundary condition $x^o(t_b) = c$ as follows:

$$\dot{x}^o(t) = ax^o(t) - b^2 e^{-at}\lambda^o(0)$$

$$x^o(t) = e^{at}x_0 - \int_0^t e^{a(t-\sigma)}\left(b^2 e^{-a\sigma}\lambda^o(0)\right) d\sigma$$

$$= e^{at}x_0 - b^2\lambda^o(0)e^{at}\int_0^t e^{-2a\sigma} d\sigma$$

$$= e^{at}x_0 + \frac{b^2\lambda^o(0)}{2a}e^{at}\left(e^{-2at} - 1\right)$$

$$x^o(t_b) = c = e^{at_b}x_0 - \frac{b^2\lambda^o(0)}{a}\sinh(at_b)$$

$$\lambda^o(0) = \frac{a(e^{at_b}x_0 - c)}{b^2 \sinh(at_b)} > 0 .$$

Therefore, the explicit formula for the optimal open-loop control is:

$$u^o(t) = -b\lambda^o(t) = -\frac{a(e^{at_b}x_0 - c)}{b \sinh(at_b)}e^{-at} .$$

2.5 Optimal Control Problems with State Constraints

In this section, Pontryagin's Minimum Principle is derived for optimal control problems of the general form of Type B, but with the additional state constraint $x^o(t) \in \Omega_x(t)$ for all $t \in [t_a, t_b]$ for some closed set $\Omega_x(t) \subset R^n$.

As an example, the time-optimal control problem for the horizontal, frictionless motion of a mass point with a velocity constraint is solved.

2.5.1 The Optimal Control Problem of Type D

Statement of the optimal control problem:

Find a piecewise continuous control $u^o : [t_a, t_b] \to \Omega \subseteq R^m$, such that the constraints

$$x^o(t_a) = x_a$$

$$\dot{x}^o(t) = f(x^o(t), u^o(t), t) \quad \text{for all } t \in [t_a, t_b]$$

$$x^o(t) \in \Omega_x(t) \quad \text{for all } t \in [t_a, t_b] ,$$
$$\Omega_x(t) = \{x \in R^n \mid G(x, t) \leq 0; \ G : R^n \times [t_a, t_b] \to R\}$$

$$x^o(t_b) \in S \subseteq R^n$$

are satisfied and such that the cost functional

$$J(u) = K(x^o(t_b), t_b) + \int_{t_a}^{t_b} L(x^o(t), u^o(t), t)\, dt$$

is minimized;

Subproblem D.1: t_b is fixed,

Subproblem D.2: t_b is free $(t_b > t_a)$.

Remark: t_a, $x_a \in R^n$, $\Omega_x(t) \subset R^n$, and $S \subseteq R^n$ are specified; $\Omega \subseteq R^m$ is time-invariant. The state constraint $\Omega_x(t)$ is defined by the scalar inequality $G(x, t) \leq 0$. The function $G(x, t)$ is assumed to be continuously differentiable. Of course, the state constraint could also be described by several inequalities, each of which could be active or inactive at any given time t.

2.5.2 Pontryagin's Minimum Principle

For the sake of simplicity, it is assumed that the optimal control problem is regular with $\lambda_0^o = 1$. Thus, the Hamiltonian function is

$$H(x(t), u(t), \lambda(t), t) = L(x(t), u(t), t) + \lambda^{\mathrm{T}}(t) f(x(t), u(t), t)\ .$$

Assumption:

In the formulation of Theorem D below, it is assumed that the state constraint $x^o(t) \in \Omega_x(t)$ is active in a subinterval $[t_1, t_2]$ of $[t_a, t_b]$ and inactive for $t_a \leq t < t_1$ and $t_2 < t \leq t_b$.

The following notation for the function G and its total derivatives with respect to time along an optimal trajectory is used:

$$G^{(0)}(x(t), t) = G(x(t), t)$$

$$G^{(1)}(x(t), t) = \frac{d}{dt} G(x(t), t) = \frac{\partial G(x(t), t)}{\partial x} \dot{x}(t) + \frac{\partial G(x(t), t)}{\partial t}$$

$$G^{(2)}(x(t), t) = \frac{d}{dt} G^{(1)}(x(t), t)$$

$$\vdots$$

$$G^{(\ell-1)}(x(t), t) = \frac{d}{dt} G^{(\ell-2)}(x(t), t)$$

$$G^{(\ell)}(x(t), u(t), t) = \frac{d}{dt} G^{(\ell-1)}(x(t), t)$$

Note: In $G^{(\ell)}$, u appears explicitly for the first time. Obviously, $\ell \geq 1$.

Theorem D

If the control $u^o : [t_a, t_b] \rightarrow \Omega$ is optimal (in the non-singular case with $\lambda_0^o = 1$), then the following conditions are satisfied:

a) $\dot{x}^o(t) = \nabla_\lambda H_{|o} = f(x^o(t), u^o(t), t)$

   for $t \notin [t_1, t_2]$:
   $$\dot{\lambda}^o(t) = -\nabla_x H_{|o}$$
   $$= -\nabla_x L(x^o(t), u^o(t), t) - \left[\frac{\partial f}{\partial x}(x^o(t), u^o(t), t) \right]^T \lambda^o(t)$$

   for $t \in [t_1, t_2]$:
   $$\dot{\lambda}^o(t) = -\nabla_x \overline{H}_{|o} = -\nabla_x H_{|o} - \mu_\ell^o(t) \nabla_x G^{(\ell)}{}_{|o}$$
   $$= -\nabla_x L(x^o(t), u^o(t), t) - \left[\frac{\partial f}{\partial x}(x^o(t), u^o(t), t) \right]^T \lambda^o(t)$$
   $$- \mu_\ell^o(t) \nabla_x G^{(\ell)}(x^o(t), u^o(t), t)$$

   with $\mu_\ell^o(t) \geq 0$

$x^o(t_a) = x_a$

   for $t \notin [t_1, t_2]$: $x^o(t) \in \text{int}(\Omega_x)$, i.e., $G(x^o(t), t) < 0$

   for $t \in [t_1, t_2]$: $x^o(t) \in \partial\Omega_x$, i.e., $G(x^o(t), t) \equiv 0$,
   in particular:
   for $t = t_2$ (or equivalently for $t = t_1$):
   $G(x^o(t), t) = G^{(1)}(x^o(t), t) = \cdots = G^{(\ell-1)}(x^o(t), t) = 0$
   and
   for $t \in [t_1, t_2]$: $G^{(\ell)}(x^o(t), u^o(t), t) \equiv 0$

$x^o(t_b) \in S$

   for $t = t_2$ (or alternatively for $t = t_1$):
   $$\lambda^o(t_{2-}) = \lambda^o(t_{2+}) + \sum_{0}^{\ell-1} \mu_i^o \nabla_x G^{(i)}(x^o(t_2), t_2)$$
   with $\mu_i^o \geq 0$ for all i, $i = 0, \ldots, \ell - 1$

   $\lambda^o(t_b) = \nabla_x K(x^o(t_b), t_b) + q^o$ with $q^o \in T^*(S, x^o(t_b))^2$.

b) For $t \notin [t_1, t_2]$, the Hamiltonian $H(x^o(t), u, \lambda^o(t), t)$ has a global minimum with respect to $u \in \Omega$ at $u^o(t)$, i.e.,

   $H(x^o(t), u^o(t), \lambda^o(t), t) \leq H(x^o(t), u, \lambda^o(t), t)$
   for all $u \in \Omega$ and all t.

---

^{2} Normal cone of the tangent cone $T(S, x^o(t_b))$ of S at $x^o(t_b)$.

For $t \in [t_1, t_2]$, the augmented Hamiltonian function
$$\overline{H} = H(x^o(t), u, \lambda^o(t), t) + \mu_\ell^o(t) G^{(\ell)}(x^o(t), u, t)$$
has a global minimum with respect to all $\{u \in \Omega \mid G^{(\ell)}(x^o(t), u, t) = 0\}$
at $u^o(t)$, i.e.,
$$H(x^o(t), u^o(t), \lambda^o(t), t) + \mu_\ell^o(t) G^{(\ell)}(x^o(t), u^o(t), t)$$
$$\leq H(x^o(t), u, \lambda^o(t), t) + \mu_\ell^o(t) G^{(\ell)}(x^o(t), u, t)$$
for all $u \in \Omega$ with $G^{(\ell)}(x^o(t), u, t) = 0$ and all $t \in [t_1, t_2]$.

c) Furthermore, if the final time t_b is free (Subproblem D.2):
$$H(x^o(t_b), u^o(t_b), \lambda^o(t_b), t_b) = -\frac{\partial K}{\partial t}(x^o(t_b), t_b) \ .$$

2.5.3 Proof

Essentially, proving Theorem D proceeds in complete analogy to the proofs of Theorems A and B given in Chapters 2.1.3 and 2.3.3, respectively. However, there is the minor complication that the optimal state $x^o(.)$ has to slide along the boundary of the state constraint set $\Omega_x(t)$ in the interval $t \in [t_1, t_2]$ as assumed in the formulation of Theorem D.

In keeping with the principle of optimality (see Chapter 3.1), the author prefers to handle these minor complications as follows:

At time t_2, the following conditions must hold simultaneously:
$$G(x^o(t_2), t_2) = 0$$
$$G^{(1)}(x^o(t_2), t_2) = 0$$
$$\vdots$$
$$G^{(\ell-1)}(x^o(t_2), t_2) = 0 \ .$$
Furthermore, the condition
$$G^\ell(x^o(t), t) = 0$$
must be satisfied for all $t \in [t_1, t_2]$.

Of course, alternatively and equivalently, the ℓ conditions for G, $G^{(1)}$, ..., $G^{(\ell-1)}$ could be stated for $t = t_1$ rather than for $t = t_2$.

With the Lagrange multiplier vectors λ_a and $\lambda(t)$ for $t \in [t_a, t_b]$ and the scalar Lagrange multipliers μ_0, μ_1, ..., $\mu_{\ell-1}$, and $\mu_\ell(t)$ for $t \in [t_1, t_2]$, the augmented cost functional $\overline{J}$ can be written in the following form:

$$\overline{J} = K(x(t_b), t_b) + \int_{t_a}^{t_b} \left[L(x, u, t) + \lambda(t)^T \{f(x, u, t) - \dot{x}\}\right] dt$$

$$+ \lambda_a^T \{x_a - x(t_a)\} + \int_{t_1}^{t_2} \mu_\ell(t) G^{(\ell)}(x, u, t) \, dt + \sum_{i=0}^{\ell-1} \mu_i G^{(i)}(x(t_2), t_2) \ .$$

Defining the support function

$$\kappa(t_1, t_2) = \begin{cases} 0 & \text{for } t < t_1 \\ 1 & \text{for } t \in [t_1, t_2] \\ 0 & \text{for } t > t_2 \end{cases}$$

and using the Dirac function $\delta(\cdot)$ allows us to incorporate the last two terms of $\overline{J}$ into the first integral. The result is:

$$\overline{J} = K(x(t_b), t_b) + \lambda_a^{\mathrm{T}}\{x_a - x(t_a)\}$$

$$+ \int_{t_a}^{t_b} \Big[L(x, u, t) + \lambda^{\mathrm{T}}\{f(x, u, t) - \dot{x}\}$$

$$+ \kappa(t_1, t_2)\mu_\ell G^{(\ell)}(x, u, t) + \delta(t - t_2) \sum_{i=0}^{\ell-1} \mu_i G^{(i)}(x(t_2), t_2) \Big] \, dt$$

$$= K(x(t_b), t_b) + \lambda_a^{\mathrm{T}}\{x_a - x(t_a)\}$$

$$+ \int_{t_a}^{t_b} \Big[\overline{H} - \lambda^{\mathrm{T}}\dot{x} + \delta(t - t_2) \sum_{i=0}^{\ell-1} \mu_i G^{(i)}(x(t_2), t_2) \Big] \, dt \;,$$

where

$$\overline{H} = L(x, u, t) + \lambda^{\mathrm{T}} f(x, u, t) + \kappa(t_1, t_2)\mu_\ell G^{(\ell)}(x, u, t)$$

$$= H(x, u, \lambda, t) + \kappa(t_1, t_2)\mu_\ell G^{(\ell)}(x, u, t) \;.$$

According to the philosophy of the Lagrange multiplier method, the augmented cost functional $\overline{J}$ has to be minimized with respect to all of its mutually independent variables $x(t_a)$, λ_a, $x(t_b)$, and $u(t)$, $x(t)$, and $\lambda(t)$ for all $t \in (t_a, t_b)$, and $\mu_0, \dots \mu_{\ell-1}$, and $\mu_\ell(t)$ for all $t \in [t_1, t_2]$, as well as t_b (if the final time is free). — Note that all of the scalar Lagrange multipliers μ_i must be non-negative (see Chapter 1.3.2).

Suppose, we have found the optimal solution $x^o(t_a)$, λ_a^o, $x^o(t_b)$ (satisfying $x^o(t_b) \in S$), and $u^o(t)$ (satisfying $u^o(t) \in \Omega$), $x^o(t)$ (satisfying $x^o(t) \in \partial\Omega_x(t)$ for $t \in [t_1, t_2]$ and $x^o(t) \in \text{int}\{\Omega_x(t)\}$ for $t \notin [t_1, t_2]$), and $\lambda^o(t)$ for all $t \in (t_a, t_b)$, and $\mu_0^o \geq 0, \dots, \mu_{\ell-1}^o \geq 0$, and $\mu_\ell^o(t) \geq 0$ for all $t \in [t_1, t_2]$, as well as t_b (if the final time is free).

The following first differential $\delta\overline{J}$ of $\overline{J}(u^o)$ around the optimal solution is obtained (for details of the analysis, consult Chapter 2.1.3):

$$\delta\overline{J} = \left[\left(\frac{\partial K}{\partial x} - \lambda^{\mathrm{T}} \right)(\delta x + \dot{x}\delta t_b) \right]_{t_b}$$

$$+ \left[\frac{\partial K}{\partial t} + H \right]_{t_b} \delta t_b \; + \; \sum_{i=0}^{\ell-1} \delta\mu_i G^{(i)}(x(t_2), t_2)$$

$$+ \delta\lambda_a^{\mathrm{T}}\{x_a - x(t_a)\} + \left(\lambda^{\mathrm{T}}(t_a) - \lambda_a^{\mathrm{T}}\right)\delta x(t_a)$$

$$+ \int_{t_a}^{t_b}\left[\left(\frac{\partial\overline{H}}{\partial x} + \dot\lambda^{\mathrm{T}} + \delta(t-t_2)\sum_{i=0}^{\ell-1}\mu_i\frac{\partial G^{(i)}}{\partial x}\right)\delta x + \frac{\partial\overline{H}}{\partial u}\delta u\right.$$

$$\left. + \left(\frac{\partial\overline{H}}{\partial\lambda} - \dot x^{\mathrm{T}}\right)\delta\lambda + \kappa(t_1,t_2)G^{(\ell)}(x,u,t)\delta\mu_\ell\right]dt \ .$$

Since we have postulated a minimum of the augmented function at $\overline{J}(u^o)$, this first differential must satisfy the inequality

$$\delta\overline{J} \geq 0$$

for all admissible variations of the independent variables. All of the variations of the independent variables are unconstrained, with the exceptions that $\delta u(t)$ is constrained to the tangent cone of Ω at $u^o(t)$, i.e.,

$$\delta u(t) \in T(\Omega, u^o(t)) \quad \text{for all } t \in [t_a, t_b] \ ,$$

such that the control constraint $u(t) \in \Omega$ is not violated, and that $\delta x(t_b)$ is constrained to the tangent cone of S at $x^o(t_b)$, i.e.,

$$\delta x(t_b) \in T(S, x^o(t_b)) \ ,$$

such that the constraint $x(t_b) \in S$ is not violated, and

$$\delta t_b = 0$$

if the final time is fixed (Problem Type D.1).

According to the philosophy of the Lagrange multiplier method, this inequality must hold for arbitrary combinations of the mutually independent variations δt_b, and $\delta x(t)$, $\delta u(t)$, $\delta\lambda(t)$, $\delta\mu_\ell(t)$ at any time $t \in (t_a, t_b)$, and $\delta\lambda_a$, $\delta x(t_a)$, and $\delta x(t_b)$. Therefore, this inequality must be satisfied for a few very specially chosen combinations of these variations as well, namely where only one single variation is nontrivial and all of the others vanish.

The consequence is that all of the factors multiplying a differential must vanish.

There are three exceptions:

1) If the final time is fixed, it must not be varied; therefore, the second bracketed term must only vanish if the final time is free.

2) If the optimal control $u^o(t)$ at time t lies in the interior of the control constraint set Ω, then the factor $\partial H/\partial u$ must vanish (and H must have a

local minimum). If the optimal control $u^o(t)$ at time t lies on the boundary $\partial\Omega$ of Ω, then the inequality must hold for all $\delta u(t) \in T(\Omega, u^o(t))$. However, the gradient $\nabla_u H$ need not vanish. Rather, $-\nabla_u H$ is restricted to lie in the normal cone $T^*(\Omega, u^o(t))$, i.e., again, the Hamiltonian must have a (local) minimum at $u^o(t)$.

3) If the optimal final state $x^o(t_b)$ lies in the interior of the target set S, then the factor in the first round brackets must vanish. If the optimal final state $x^o(t_b)$ lies on the boundary ∂S of S, then the inequality must hold for all $\delta x(t_b) \in T(S, x^o(t_b))$. In other words, $\lambda^o(t_b)$ can be of the form

$$\lambda^o(t_b) = \lambda_0^o \nabla_x K(x^o(t_b), t_b) + q \ ,$$

where q must lie in the normal cone $T^*(S, x^o(t_b))$ of the target set S at $x^o(t_b)$. This guarantees that the resulting term satisfies

$$-q^{\mathrm{T}} \delta x(t_b) \geq 0$$

for all permissible variations $\delta x(t_b)$ of the final state $x^o(t_b)$.

This completes the proof of Theorem D.

Notice that there is no condition for λ_a. In other words, the boundary condition $\lambda^o(t_a)$ of the optimal costate $\lambda^o(.)$ is free.

Remarks:

- The calculus of variations only requests the local minimization of the Hamiltonian $\overline{H}$ with respect to the control u. — In Theorem D, the Hamiltonian $\overline{H}$ is requested to be globally minimized over the admissible set Ω. This restriction is justified in Chapter 2.2.1.

- Note that the expression for the differential $\delta\overline{J}$ contains none of the variations $\delta\mu_0, \ldots, \delta\mu_{\ell-1}, \delta\mu_\ell(t), \delta t_1$, and δt_2 because all of them are multiplied by vanishing factors.

2.5.4 Time-Optimal, Frictionless, Horizontal Motion of a Mass Point with a Velocity Constraint

The problem considered in this section is almost identical to Problem 1 (Chapter 1.2, p. 5) which has been solved in Chapter 2.1.4.1, but now, the the velocity constraint $|x_2(t)| \leq 1$ must be obeyed at all times $t \in [0, t_b]$.

Statement of the optimal control problem:

Find $u : [0, t_b] \to [-1, 1]$, such that the dynamic system

$$\dot{x}_1(t) = x_2(t)$$
$$\dot{x}_2(t) = u(t)$$

with the velocity constraint

$$|x_2(t)| \leq 1 \quad \text{for } t \in [0, t_b]$$

is transferred from the initial state

$$x_1(0) = 10$$
$$x_2(0) = 0$$

to the final state

$$x_1(t_b) = 0$$
$$x_2(t_b) = 0$$

and such that the cost functional

$$J(u) = \int_0^{t_b} dt$$

is minimized.

Remark: For the analysis of this problem, the velocity constraint is best described in the following form:

$$G(x_1, x_2) = x_2^2 - 1 \leq 0 \ .$$

The optimal solution:

Obviously, the optimal solution is characterized as follows:

$t = 0$		$x_2^o = 0$	$x_1^o = 10$
$t = 0 \ldots 1$	$u^o \equiv -1$		
$t = t_1 = 1$		$x_2^o = -1$	$x_1^o = 9.5$
$t = 1 \ldots 10$	$u^o \equiv 0$	$x_2^o \equiv -1$	
$t = t_2 = 10$		$x_2^o = -1$	$x_1^o = 0.5$
$t = 10 \ldots 11$	$u^o \equiv 1$		
$t_b = 11$		$x_2^o = 0$	$x_1^o = 0 \ .$

Intermezzo:

The optimal control problem is regular, i.e., $\lambda_0^o = 1$.

For formulating Pontryagin's necessary conditions for the optimality of a solution, the following items are needed:

$$G(x_1, x_2) = x_2^2 - 1$$
$$\dot{G}(x_1, x_2) = 2x_2 u \ , \quad \text{hence} \quad \ell = 1$$
$$\nabla_x G(x_1, x_2) = \begin{bmatrix} 0 \\ 2x_2 \end{bmatrix}$$
$$\nabla_x \dot{G}(x_1, x_2) = \begin{bmatrix} 0 \\ 2u \end{bmatrix} \ .$$

The Hamiltonian functions are

$$H = 1 + \lambda_1 x_2 + \lambda_2 u$$

and

$$\overline{H} = 1 + \lambda_1 x_2 + \lambda_2 u + \kappa(1, 10) 2\mu_1(t) x_2 u \ ,$$

respectively.

Pontryagin's necessary conditions:

a) $\dot{x}_1 = x_2$

   $\dot{x}_2 = u \qquad u \in [-1, +1]$

   for $t \notin [1, 10]$:

   $$\dot{\lambda}_1 = -\frac{\partial H}{\partial x_1} = 0$$

   $$\dot{\lambda}_2 = -\frac{\partial H}{\partial x_2} = -\lambda_1$$

   for $t \in [1, 10]$:

   $$\dot{\lambda}_1 = -\frac{\partial \overline{H}}{\partial x_1} = 0$$

   $$\dot{\lambda}_2 = -\frac{\partial \overline{H}}{\partial x_2} = -\lambda_1 - \mu_1(t) 2u \qquad \mu_1(t) \geq 0$$

   $$\dot{G}(x, u) = 2x_2 u \equiv 0$$

   $x_1(0) = 10$
   $x_2(0) = 0$

   for $t = 10$:

   $G(x) = x_2^2(10) - 1 = 0$
   $\lambda_2(10^-) = \lambda_2(10^+) + \mu_0 2x_2(10) \qquad \mu_0 \geq 0$

   $x_1(11) = 0$
   $x_2(11) = 0$

b) for $t \notin [1, 10]$:
    minimize the Hamiltonian H w.r. to $u \in [-1, +1]$

    for $t \in [1, 10]$:
    minimize the Hamiltonian $\overline{H}$ w.r. to $u \in [-1, +1]$
    under the constraint $\dot{G}(x, u) = 2x_2 u = 0$

c) $H \equiv H(11) = 0 \,,$
    since the problem is time-invariant and the final time t_b is free.

The minimization of the Hamiltonian functions yields the following results:

For $t \in [0, 1)$ and $t \in (10, 11]$ (see Chapter 2.1.4.1):

$$u^o(t) = -\text{sign}\{\lambda_2^o(t)\} = \begin{cases} +1 & \text{for } \lambda_2^o(t) < 0 \\ 0 & \text{for } \lambda_2^o(t) = 0 \\ -1 & \text{for } \lambda_2^o(t) > 0 \ . \end{cases}$$

For $t \in [1, 10]$, the scalar optimal control u is determined by the constraint $\dot{G} \equiv 0$ alone and hence

$$u^o(t) \equiv 0 \quad \text{while} \quad x_2^o(t) \equiv -1 \ .$$

It remains to be proved that the dynamics of the costate variables $\lambda_1(t)$ and $\lambda_2(t)$ admit a solution leading to the proclaimed optimal control $u^o(.)$.

Since $\dot{\lambda}_1^0(t)$ vanishes at all times, the costate variable $\lambda_1^o(t)$ is constant:

$$\lambda_1^o(t) \equiv \lambda_1^o \ .$$

In the interval $t \in [1, 10]$, the control u vanishes. Therefore, $\lambda_2^o(t)$ is an affine function of the time t. Since we are working our way backwards in time, we can write:

$$\lambda_2^o(t) = \lambda_2^o(11) + \lambda_1^o(11 - t) \ ,$$

where, of course, the values of λ_1^o and $\lambda_2^o(11)$ are unknown yet.

In order for the proposed control $u^o(.)$ to be optimal we need the following conditions:

$$\lambda_2^o(t) = \begin{cases} \lambda_2^o(11) < 0 & \text{for } t = 11 \\ \lambda_2^o(11) + \lambda_1^o \leq 0 & \text{for } t = 10^+ \\ \lambda_2^o(11) + \lambda_1^o - 2\mu_0^o & \text{for } t = 10^-, \ \mu_o^o \geq 0 \\ \lambda_2^o(11) + 10\lambda_1^o - 2\mu_0^o \geq 0 & \text{for } t = 1 \\ \lambda_2^o(11) + 11\lambda_1^o - 2\mu_0^o > 0 & \text{for } t = 0 \ . \end{cases}$$

Thus, at the moment, we have three unknowns: λ_1^o, $\lambda_2^o(11)$, and μ_0^o.

Exploiting condition c of Pontryagin's Minimum Principle,

$$H|_o = 1 + \lambda_1^o x_2^o(t) + \lambda_2^o(t) u^o(t) \equiv 0 \ ,$$

and again working backwards in time evolves in the following steps:

At the final time $t_b = 11$:

$$H(11) = 0 = 1 + \lambda_1^o \cdot 0 + \lambda_2^o(11) \cdot 1 \ .$$

Thus, we find $\lambda_2^o(11) = -1$.

At $t = 10^+$:

$$H(10^+) = 0 = 1 + \lambda_1^o \cdot (-1) + (-1 + \lambda_1^o) \cdot (+1) \,.$$

At $t = 10^-$:

$$H(10^-) = 0 = 1 + \lambda_1^o \cdot (-1) + (-1 + \lambda_1^o - 2\mu_0^o) \cdot 0 \,.$$

This condition yields the result $\lambda_1^o = 1$.

At time $t = 1^+$:

$$H(1^+) = 0 = 1 + 1 \cdot (-1) + (9 - 2\mu_0^o) \cdot 0 \,.$$

At time $t = 1^-$:

$$H(1^-) = 0 = 1 + 1 \cdot (-1) + (9 - 2\mu_0^o) \cdot (-1) \,.$$

Therefore, we must have $\mu_0^o = 4.5$.

And finally, we verify that the Hamiltonian vanishes at the initial time $t = 0$ as well:

$$H(0) = 0 = 1 + 1 \cdot 0 + 1 \cdot (-1) \,.$$

In summary, we have found:

$$\lambda_2^o(t) = \begin{cases} -(t - 10) & \text{for } t = 10^+ \ldots 11 \\ -(t - 1) & \text{for } t = 0 \ldots 10^- \end{cases}$$

and

$$u^o(t) = \begin{cases} +1 & \text{for } t = 10^+ \ldots 11 \\ 0 & \text{for } t = 1^+ \ldots 10^- \\ -1 & \text{for } t = 0 \ldots 1^- \,. \end{cases}$$

All of the necessary conditions of Pontryagin's Minimum Principle are satisfied. And we have found the optimal solution, indeed.

2.6 Singular Optimal Control

In this section, the special case is treated where the Hamiltonian function H is not an explicit function of the control u during a time interval $[t_1, t_2] \subseteq [t_a, t_b]$. Therefore, the optimal control cannot be found directly by globally minimizing H. — This special case has already been mentioned in the introductory text of Chapter 2 on p. 23.

Two optimal control problems involving singular optimal controls are analyzed in Chapters 2.6.2 and 2.6.3.

2.6.1 Problem Solving Technique

Most often, this special case arises if the Hamiltonian function H is an affine function of the control u. For the sake of simplicity, the discussion concentrates on a time-invariant optimal control problem with a constrained scalar control variable.

In this case, the Hamiltonian function has the following form:

$$H = g(x(t), \lambda(t), \lambda_0) + h(x(t), \lambda(t), \lambda_0)u(t)$$

with $u \in [u_{\min}, u_{\max}] \subset R$ and where f and g are two scalar-valued functions: $f : R^n \times R^n \times \{0, 1\} \to R$, $g : R^n \times R^n \times \{0, 1\} \to R$.

The singular case occurs when the "switching function" h vanishes at all times t in a time interval $[t_1, t_2]$:

$$h(x^o(t), \lambda^o(t), \lambda_0^o) \equiv 0 \quad \text{for } t \in [t_1, t_2] \subseteq [t_a, t_b] .$$

Obviously, in this situation, all $u \in [u_{\min}, u_{\max}]$ globally minimize the Hamiltonian function H.

A work-around for this nasty situation follows from the fact that, with $h \equiv 0$ in this time interval, all of the total derivatives of the switching function h along the optimal trajectory must vanish in this time interval as well, i.e., $\dot{h} \equiv 0$, $\ddot{h} \equiv 0$, $h^{(3)} \equiv 0$, and so on.

The differentiation is continued until the control variable u explicitly appears in a derivative. Interestingly, this always happens in an even derivative $h^{(2k)} \equiv 0$ for some $k \geq 1$. (See [24] for a proof of this fact and [22] for some interesting comments.)

Thus, we obtain the following necessary conditions for an optimal solution to evolve along a singular arc in the time interval $[t_1, t_2]$:

$$h^{(2k)}(x^o(t), u^o(t), \lambda^o(t), \lambda_0^o) \equiv 0 , \text{ and}$$

$$h^{(2k-1)}(x^o(t), \lambda^o(t), \lambda_0^o) \equiv 0 , \text{ and}$$

$$\vdots$$

$$\vdots$$

$$\dot{h}(x^o(t), \lambda^o(t), \lambda_0^o) \equiv 0 \text{ , and}$$

$$h(x^o(t), \lambda^o(t), \lambda_0^o) \equiv 0 \text{ .}$$

If the final time t_b is free, then the condition

$$H = g(x^o(t), \lambda^o(t), \lambda_0^o) \equiv 0$$

must be satisfied as well.

In order to determine the singular optimal control, we proceed as follows: First we solve the condition $h^{(2k)} \equiv 0$ for $u^o(t)$. Then, we verify that the corresponding trajectories $x^o(.)$ and $\lambda^o(.)$ satisfy the additional $2k$ conditions $h^{(2k-1)} \equiv 0$, ..., $\dot{h} \equiv 0$, and $h \equiv 0$ (and $H \equiv 0$ if the final time t_b is free) simultaneously as well.

If all of these necessary conditions for a singular optimal arc are satisfied, we have found a candidate for an optimal control. Note that no sufficiency condition is implied.

2.6.2 Goh's Fishing Problem

Statement of the optimal control problem:

See Chapter 1.2, Problem 6, p. 10. — If the catching capacity U of the fishing fleet and the final time t_b are sufficiently large, the optimal solution contains a singular arc.

Hamiltonian function:

$$H = -\lambda_0 u(t) + \lambda(t) a x(t) - \lambda(t) \frac{a}{b} x^2(t) - \lambda(t) u(t) \text{ .}$$

Pontryagin's necessary conditions for optimality:

If $u^o : [0, t_b] \rightarrow [0, U]$ is an optimal control, then there exists a nontrivial vector

$$\begin{bmatrix} \lambda_0^o \\ \lambda^o(t_b) \end{bmatrix} \neq \begin{bmatrix} 0 \\ 0 \end{bmatrix} ,$$

such that the following conditions are satisfied:

a) Differential equations and boundary conditions:

$$\dot{x}^o(t) = a x^o(t) - \frac{a}{b} x^{o2}(t) - u^o(t)$$

$$x^o(t) \geq 0 \quad \text{for all } t \in [0, t_b]$$

$$\dot{\lambda}^o(t) = -\frac{\partial H}{\partial x} = a\left(\frac{2}{b}x^o(t) - 1\right)\lambda^o(t)$$

$$x^o(0) = x_a > 0$$

$$x^o(t_b) \geq 0$$

$$\lambda^o(t_b) \begin{cases} = 0 & \text{for } x^o(t_b) > 0 \\ \leq 0 & \text{for } x^o(t_b) = 0 \ . \end{cases}$$

b) Minimization of the Hamiltonian function:

$$H(x^o(t), u^o(t), \lambda^o(t), \lambda_0^o) \leq H(x^o(t), u, \lambda^o(t), \lambda_0^o)$$
$$\text{for all } u \in [0, U] \text{ and all } t \in [0, t_b]$$

and hence

$$-\lambda_0^o u^o(t) - \lambda^o(t)u^o(t) \leq -\lambda_0^o u - \lambda^o(t)u$$
$$\text{for all } u \in [0, U] \text{ and all } t \in [0, t_b] \ .$$

For convenience, we write the Hamiltonian function in the form

$$H = g(x(t), \lambda(t), \lambda_0) - h(x(t), \lambda(t), \lambda_0)u(t)$$

with the switching function

$$h(x(t), \lambda(t), \lambda_0) = \lambda_0^o + \lambda^o(t) \ .$$

Minimizing the Hamiltonian function yields the following preliminary control law:

$$u^o(t) = \begin{cases} U & \text{for } h(t) > 0 \\ u \in [0, U] & \text{for } h(t) = 0 \\ 0 & \text{for } h(t) < 0 \ . \end{cases}$$

Analysis of a potential singular arc:

If there is a singular arc, the switching function h and its first and second total derivate $\dot{h}$ and $\ddot{h}$, respectively, have to vanish simultaneously along the corresponding trajectories $x(.)$ and $\lambda(.)$, i.e.:

$$h = \lambda_0 + \lambda \equiv 0$$

$$\dot{h} = \dot{\lambda} = a\left(\frac{2}{b}x - 1\right)\lambda \equiv 0$$

$$\ddot{h} = \frac{2a}{b}\dot{x}\lambda + a\left(\frac{2}{b}x - 1\right)\dot{\lambda}$$

$$= \frac{2a}{b}\left(ax - \frac{a}{b}x^2 - u\right)\lambda + a^2\left(\frac{2}{b}x - 1\right)^2\lambda \equiv 0 \ .$$

Due to the nontriviality requirement for the vector $(\lambda_0, \lambda(t_b))$, the following conditions must be satisfied on a singular arc:

$$\lambda_0^o = 1$$

$$\lambda^o(t) \equiv -1$$

$$x^o(t) \equiv \frac{b}{2}$$

$$u^o(t) \equiv \frac{ab}{4} \leq U \; .$$

Therefore, a singular arc is possible, if the catching capacity U of the fleet is sufficiently large, namely $U \geq ab/4$. — Note that both the fish population $x(t)$ and the catching rate $u(t)$ are constant on the singular arc.

Since the differential equation governing $\lambda(t)$ is homogeneous, an optimal singular arc can only occur, if the fish population is exterminated (exactly) at the final time t_b with $\lambda(t_b) = -1$, because otherwise $\lambda(t_b)$ would have to vanish.

An optimal singular arc occurs, if the initial fish population x_a is sufficiently large, such that $x = b/2$ can be reached. Obviously, the singular arc can last for a very long time, if the final time t_b is very large. — This is the sustainability aspect of this dynamic system. Note that the singular "equilibrium" of this nonlinear system is semi-stable.

The optimal singular arc begins when the population $x = b/2$ is reached (either from above with $u^o(t) \equiv U$ or from below with $u^o(t) \equiv 0$. It ends when it becomes "necessary" to exterminate the fish exactly at the final time t_b by applying $u^o(t) \equiv U$.

For more details about this fascinating problem, see [18].

2.6.3 Fuel-Optimal Atmospheric Flight of a Rocket

Statement of the optimal control problem:

See Chapter 1.2, Problem 4, p. 7. — The problem has a (unique) optimal solution, provided the specified final state x_b lies in the set of states which are reachable from the given initial state x_a at the fixed final time t_b. If the final state lies in the interior of this set, the optimal solution contains a singular arc where the rocket flies at a constant speed. ("Kill as much time as possible while flying at the lowest possible constant speed.")

Minimizing the fuel consumption $\int_0^{t_b} u(t)dt$ is equivalent to maximizing the final mass $x_3(t_b)$ of the rocket. Thus, the most suitable cost functional is

$$J(u) = x_3(t_b) \; ,$$

which we want to maximize. — It has the special form which has been discussed in Chapter 2.2.3. Therefore, $\lambda_3(t_b)$ takes over the role of λ_0. Since we want to maximize the cost functional, we use Pontryagin's Maximum Principle, where the Hamiltonian has to be globally maximized (rather than minimized).

Hamiltonian function:

$$H = \lambda_1 \dot{x}_1 + \lambda_2 \dot{x}_2 + \lambda_3 \dot{x}_3$$

$$= \lambda_1 x_2 + \frac{\lambda_2}{x_3}\left(u - \frac{1}{2} A\rho c_w x_2^2\right) - \alpha\lambda_3 u \ .$$

Pontryagin's necessary conditions for optimality:

If $u^o : [0, t_b] \rightarrow [0, F_{\max}]$ is an optimal control, then there exists a nontrivial vector

$$\begin{bmatrix} \lambda_1^o(t_b) \\ \lambda_2^o(t_b) \\ \lambda_3^o(t_b) \end{bmatrix} \neq \begin{bmatrix} 0 \\ 0 \\ 0 \end{bmatrix} \text{ with } \lambda_3^o(t_b) = \begin{cases} 1 & \text{in the regular case} \\ 0 & \text{in a singular case,} \end{cases}$$

such that the following conditions are satisfied:

a) Differential equations and boundary conditions:

$$\dot{x}_1^o(t) = x_2^o(t)$$

$$\dot{x}_2^o(t) = \frac{1}{x_3^o(t)}\left(u^o(t) - \frac{1}{2} A\rho c_w x_2^{o2}(t)\right)$$

$$\dot{x}_3^o(t) = -\alpha u^o(t)$$

$$\dot{\lambda}_1^o(t) = -\frac{\partial H}{\partial x_1} = 0$$

$$\dot{\lambda}_2^o(t) = -\frac{\partial H}{\partial x_2} = -\lambda_1^o(t) + A\rho c_w \frac{\lambda_2^o(t) x_2^o(t)}{x_3^o(t)}$$

$$\dot{\lambda}_3^o(t) = -\frac{\partial H}{\partial x_3} = \frac{\lambda_2^o(t)}{x_3^{o2}(t)}\left(u^o(t) - \frac{1}{2} A\rho c_w x_2^{o2}(t)\right) \ .$$

b) Maximization of the Hamiltonian function:

$$H(x^o(t), u^o(t), \lambda^o(t)) \geq H(x^o(t), u, \lambda^o(t))$$

$$\text{for all } u \in [0, F_{\max}] \text{ and all } t \in [0, t_b]$$

and hence

$$\left(\frac{\lambda_2^o(t)}{x_3^o(t)} - \alpha\lambda_3^o(t)\right) u^o(t) \geq \left(\frac{\lambda_2^o(t)}{x_3^o(t)} - \alpha\lambda_3^o(t)\right) u$$

$$\text{for all } u \in [0, F_{\max}] \text{ and all } t \in [0, t_b] \ .$$

With the switching function

$$h(t) = \frac{\lambda_2^o(t)}{x_3^o(t)} - \alpha\lambda_3^o(t),$$

maximizing the Hamiltonian function yields the following preliminary control law:

$$u^o(t) = \begin{cases} F_{\max} & \text{for } h(t) > 0 \\ u \in [0, F_{\max}] & \text{for } h(t) = 0 \\ 0 & \text{for } h(t) < 0 \ . \end{cases}$$

Analysis of a potential singular arc:

If there is a singular arc, the switching function h and its first and second total derivative $\dot{h}$ and $\ddot{h}$, respectively, have to vanish simultaneously along the corresponding trajectories $x(.)$ and $\lambda(.)$, i.e.:

$$h(t) = \frac{\lambda_2}{x_3} - \alpha\lambda_3 \equiv 0$$

$$\dot{h}(t) = \frac{\dot{\lambda}_2}{x_3} - \frac{\lambda_2\dot{x}_3}{x_3^2} - \alpha\dot{\lambda}_3$$

$$= -\frac{\lambda_1}{x_3} + A\rho c_w \frac{\lambda_2 x_2}{x_3^2} + \frac{\alpha\lambda_2 u}{x_3^2} - \frac{\alpha\lambda_2}{x_3^2}\left(u - \frac{1}{2}A\rho c_w x_2^2\right)$$

$$= -\frac{\lambda_1}{x_3} + A\rho c_w \frac{\lambda_2}{x_3^2}\left(x_2 + \frac{\alpha}{2}x_2^2\right)$$

$$\equiv 0$$

$$\ddot{h}(t) = -\frac{\dot{\lambda}_1}{x_3} + \frac{\lambda_1\dot{x}_3}{x_3^2} + A\rho c_w\left(\frac{\dot{\lambda}_2}{x_3^2} - \frac{2\lambda_2\dot{x}_3}{x_3^3}\right)\left(x_2 + \frac{\alpha}{2}x_2^2\right)$$

$$\quad + A\rho c_w \frac{\lambda_2}{x_3^2}(1 + \alpha x_2)\dot{x}_2$$

$$= -\frac{\alpha\lambda_1 u}{x_3^2}$$

$$\quad + A\rho c_w\left(x_2 + \frac{\alpha}{2}x_2^2\right)\left(-\frac{\lambda_1}{x_3^2} + A\rho c_w \frac{\lambda_2 x_2}{x_3^3} + \frac{2\alpha\lambda_2 u}{x_3^3}\right)$$

$$\quad + A\rho c_w \frac{\lambda_2}{x_3^3}(1 + \alpha x_2)\left(u - \frac{1}{2}A\rho c_w x_2^2\right)$$

$$\equiv 0 \ .$$

The expression for $\ddot{h}$ can be simplified dramatically by exploiting the condition $\dot{h} \equiv 0$, i.e., by replacing the terms

$$\frac{\lambda_1}{x_3^2} \quad \text{by} \quad A\rho c_w \frac{\lambda_2}{x_3^3}\left(x_2 + \frac{\alpha}{2}x_2^2\right) \ .$$

After some tedious algebraic manipulations, we get the condition

$$\ddot{h}(t) = A\rho c_w \frac{\lambda_2}{x_3^3} \left(1 + 2\alpha x_2 + \frac{\alpha^2}{2} x_2^2\right)\left(u - \frac{1}{2} A\rho c_w x_2^2\right) \equiv 0 \ .$$

Assuming that $\lambda_2(t) \equiv 0$ leads to a contradiction with Pontryagin's nontriviality condition for the vector $(\lambda_1, \lambda_2, \lambda_3)$. Therefore, $\ddot{h}$ can only vanish for the singular control

$$u^o(t) = \frac{1}{2} A\rho c_w x_2^{o2}(t) \ .$$

A close inspection of the differential equations of the state and the costate variables and of the three conditions $h \equiv 0$, $\dot{h} \equiv 0$, and $\ddot{h} \equiv 0$ reveals that the optimal singular arc has the following features:

- The velocity x_2^o and the thrust u^o are constant.

- The costate variable λ_3^o is constant.

- The ratio $\dfrac{\lambda_2^o(t)}{x_3^o(t)} = \alpha \lambda_3^o$ is constant.

- The costate variable λ_1^o is constant anyway. It attains the value
$\lambda_1^o = A\rho c_w \alpha \lambda_3^o \left(x_2^o + \dfrac{\alpha}{2} x_2^{o2}\right)$.

- If the optimal trajectory has a singular arc, then $\lambda_3^o(t_b) = 1$ is guaranteed.

We conclude that the structure of the optimal control trajectory involves three types of arcs: "boost" (where $u^o(t) \equiv F_{\max}$), "glide" (where $u^o(t) \equiv 0$), and "sustain" (corresponding to a singular arc with a constant velocity x_2).

The reader is invited to sketch all of the possible scenarios in the phase plane (x_1, x_2) and to find out what sequences of "boost", "sustain", and "glide" can occur in the optimal transfer of the rocket from (s_a, v_a) to (s_b, v_b) as the fixed final time t_b is varied from its minimal permissible value to its maximal permissible value.

2.7 Existence Theorems

One of the steps in the procedure to solve an optimal control problem is investigating whether the optimal control at hand does admit an optimal solution, indeed. — This has been mentioned in the introductory text of Chapter 2 on p. 23.

The two theorems stated below are extremely useful for the a priori investigation of the existence of an optimal control, because they cover a vast field of relevant applications. — These theorems have been proved in [26].

Theorem 1. The following optimal control problem has a globally optimal solution:

Find an unconstrained optimal control $u : [t_a, t_b] \to R^m$, such that the dynamic system
$$\dot{x}(t) = f(x(t)) + B(x(t))u(t)$$
with the continuously differentiable functions $f(x)$ and $B(x)$ is transferred from the initial state
$$x(t_a) = x_a$$
to an arbitrary final state at the fixed final time t_b and such that the cost functional
$$J(u) = K(x(t_b)) + \int_{t_a}^{t_b} \Big[L_1(x(t)) + L_2(u(t)) \Big] \, dt$$
is minimized. Here, $K(x)$ and $L_1(x)$ are convex and bounded from below and $L_2(u)$ is strictly convex and growing without bounds for all $u \in R^m$ with $\|u\| \to \infty$.

Obviously, Theorem 1 is relevant for the LQ regulator problem.

Theorem 2. Let Ω be a closed, convex, bounded, and time-invariant set in the control space R^m. — The following optimal control problem has a globally optimal solution:

Find an optimal control $u : [t_a, t_b] \to \Omega \subset R^m$, such that the dynamic system
$$\dot{x}(t) = f(x(t), u(t))$$
with the continuously differentiable function $f(x, u)$ is transferred from the initial state
$$x(t_a) = x_a$$
to an unspecified final state at the fixed final time t_b and such that the cost functional
$$J(u) = K(x(t_b)) + \int_{t_a}^{t_b} L(x(t), u(t)) \, dt$$
is minimized. Here, $K(x)$ and $L(x, u)$ are continuously differentiable functions.

Obviously, Theorem 2 can be extended to the case where the final state $x(t_b)$ at the fixed final time t_b is restricted to lie in a closed subset $S \subset R^n$, provided that the set S and the set $W(t_b) \subset R^n$ of all reachable states at the final time t_b have a non-empty intersection. — Thus, Theorem 2 covers our time-optimal and fuel-optimal control problems as well.

2.8 Optimal Control Problems with a Non-Scalar-Valued Cost Functional

Up to now, we have always considered optimal control problems with a scalar-valued cost functional. In this section, we investigate optimal control problems with non-scalar-valued cost functionals. Essentially, we proceed from the totally ordered real line $(R, \leq)$ to a partially ordered space $(\mathcal{X}_0, \preceq)$ with a higher dimension [30] into which the cost functional maps.

For a cost functional mapping into a partially ordered space, the notion of optimality splits up into superiority and non-inferiority [31]. The latter is often called Pareto optimality. Correspondingly, depending on whether we are "minimizing" or "maximizing", an extremum is called infimum or supremum for a superior solution and minimum or maximum for a non-inferior solution.

In this section, we are only interested in finding a superior solution or infimum of an optimal control problem with a non-scalar-valued cost functional.

The two most interesting examples of non-scalar-valued cost functionals are vector-valued cost functionals and matrix-valued cost functionals.

In the case of a vector-valued cost functional, we want to minimize several scalar-valued cost functionals simultaneously. A matrix-valued cost functional arises quite naturally in a problem of optimal linear filtering: We want to infimize the covariance matrix of the state estimation error. This problem is investigated in Chapter 2.8.4.

2.8.1 Introduction

Let us introduce some rather abstract notation for the finite-dimensional linear spaces, where the state $x(t)$, the control $u(t)$, and the cost $J(u)$ live:

$$\mathcal{X} : \text{state space}$$

$$\mathcal{U} : \text{input space}$$

$$\Omega \subseteq \mathcal{U} : \text{admissible set in the input space}$$

$$(\mathcal{X}_0, \preceq) : \text{cost space with the partial order } \preceq.$$

The set of all positive elements in the cost space $\mathcal{X}_0$, i.e., $\{x_0 \in \mathcal{X}_0 \mid x_0 \succeq 0\}$, is a convex cone with non-empty interior. An element $x_0 \in \mathcal{X}_0$ in the interior of the positive cone is called strictly positive: $x_0 \succ 0$.

Example: Consider the linear space of all symmetric n by n matrices which is partially ordered by positive-semidefinite difference. The closed positive cone is the set of all positive-semidefinite matrices. All elements in the interior of the positive cone are positive-definite matrices.

Furthermore, we use the following notation for the linear space of all linear maps from the linear space $\mathcal{X}$ to the linear space $\mathcal{Y}$:

$$\mathcal{L}(\mathcal{X}, \mathcal{Y}) \ .$$

Examples:

- Derivative of a function $f : R^n \to R^p : \dfrac{\partial f}{\partial x} \in \mathcal{L}(R^n, R^p)$
- Costate: $\lambda(t) \in \mathcal{L}(\mathcal{X}, \mathcal{X}_0)$
- Cost component of the extended costate: $\lambda_0 \in \mathcal{L}(\mathcal{X}_0, \mathcal{X}_0)$.

2.8.2 Problem Statement

Find a piecewise continuous control $u : [t_a, t_b] \to \Omega \subseteq \mathcal{U}$, such that the dynamic system

$$\dot{x}(t) = f(x(t), u(t), t)$$

is transferred from the initial state

$$x(t_a) = x_a$$

to an arbitrary final state at the fixed final time t_b and such that the cost

$$J(u) = K(x(t_b)) + \int_{t_a}^{t_b} L(x(t), u(t), t)\, dt$$

is infimized.

Remark: t_a, t_b, and $x_a \in \mathcal{X}$ are specified; $\Omega \subseteq \mathcal{U}$ is time-invariant.

2.8.3 Geering's Infimum Principle

Definition: Hamiltonian $H : \mathcal{X} \times \mathcal{U} \times \mathcal{L}(\mathcal{X}, \mathcal{X}_0) \times \mathcal{L}(\mathcal{X}_0, \mathcal{X}_0) \times R \to \mathcal{X}_0$,

$$H(x(t), u(t), \lambda(t), \lambda_0, t) = \lambda_0 L(x(t), u(t), t) + \lambda(t) f(x(t), u(t), t) \ .$$

Here, $\lambda_0 \in \mathcal{L}(\mathcal{X}_0, \mathcal{X}_0)$ is a positive operator, $\lambda_0 \succeq 0$. In the regular case, λ_0 is the identity operator in $\mathcal{L}(\mathcal{X}_0, \mathcal{X}_0)$, i.e., $\lambda_0 = I$.

Theorem

If $u^o : [t_a, t_b] \to \Omega$ is superior, then there exists a nontrivial pair $(\lambda_0^o, \lambda^o(t_b))$ in $\mathcal{L}(\mathcal{X}_0, \mathcal{X}_0) \times \mathcal{L}(\mathcal{X}, \mathcal{X}_0)$ with $\lambda_0 \succeq 0$, such that the following conditions are satisfied:

a) $\dot{x}^o(t) = f(x^o(t), u^o(t), t)$

$x^o(t_a) = x_a$

$\dot{\lambda}^o(t) = -\dfrac{\partial H}{\partial x}\Big|_o = -\lambda_0^o \dfrac{\partial L}{\partial x}(x^o(t), u^o(t), t) - \lambda^o(t)\dfrac{\partial f}{\partial x}(x^o(t), u^o(t), t)$

$\lambda^o(t_b) = \lambda_0^o \dfrac{\partial K}{\partial x}(x^o(t_b))$.

b) For all $t \in [t_a, t_b]$, the Hamiltonian $H(x^o(t), u, \lambda^o(t), \lambda_0^o, t)$ has a global infimum with respect to $u \in \Omega$ at $u^o(t)$, i.e.,

$H(x^o(t), u^o(t), \lambda^o(t), \lambda_0^o, t) \preceq H(x^o(t), u, \lambda^o(t), \lambda_0^o, t)$
for all $u \in \Omega$ and all $t \in [t_a, t_b]$.

Note: If we applied this notation in the case of a scalar-valued cost functional, the costate $\lambda^o(t)$ would be represented by a row vector (or, more precisely, by a 1 by n matrix).

Proof: See [12].

2.8.4 The Kalman-Bucy Filter

Consider the following stochastic linear dynamic system with the state vector $x(t) \in R^n$, the random initial state ξ, the output vector $y(t) \in R^p$, and the two white noise processes $v(t) \in R^m$ and $r(t) \in R^p$ (see [1] and [16]):

$$\dot{x}(t) = A(t)x(t) + B(t)v(t)$$
$$x(t_a) = \xi$$
$$y(t) = C(t)x(t) + r(t) .$$

The following statistical characteristics of ξ, $v(.)$, and $r(.)$ are known:

$$E\{\xi\} = x_a$$
$$E\{v(t)\} = u(t)$$
$$E\{r(t)\} = \bar{r}(t)$$
$$E\{[\xi - x_a][\xi - x_a]^T\} = \Sigma_a \geq 0$$
$$E\{[v(t) - u(t)][v(\tau) - u(\tau)]^T\} = Q(t)\delta(t - \tau) \text{ with } Q(t) \geq 0$$
$$E\{[r(t) - \bar{r}(t)][r(\tau) - \bar{r}(\tau)]^T\} = R(t)\delta(t - \tau) \text{ with } R(t) > 0 .$$

The random initial state ξ and the two white noise processes $v(.)$ and $r(.)$ are known to be mutually independent and therefore mutually uncorrelated:

$$E\{[\xi - x_a][v(\tau) - u(\tau)]^{\mathrm{T}}\} \equiv 0$$

$$E\{[\xi - x_a][r(\tau) - \overline{r}(\tau)]^{\mathrm{T}}\} \equiv 0$$

$$E\{[r(t) - \overline{r}(t)][v(\tau) - u(\tau)]^{\mathrm{T}}\} \equiv 0 \ .$$

A full-order unbiased observer for the random state vector $x(t)$ has the following generic form:

$$\dot{\widehat{x}}(t) = A(t)\widehat{x}(t) + B(t)u(t) + P(t)[y(t) - \overline{r}(t) - C(t)\widehat{x}(t)]$$

$$\widehat{x}(t_a) = x_a \ .$$

The covariance matrix $\Sigma(t)$ of the state estimation error $x(t) - \widehat{x}(t)$ satisfies the following matrix differential equation:

$$\dot{\Sigma}(t) = [A(t) - P(t)C(t)]\Sigma(t) + \Sigma(t)[A(t) - P(t)C(t)]^{\mathrm{T}}$$
$$+ B(t)Q(t)B^{\mathrm{T}}(t) + P(t)R(t)P^{\mathrm{T}}(t)$$

$$\Sigma(t_a) = \Sigma_a \ .$$

We want to find the optimal observer matrix $P^o(t)$ in the time interval $[t_a, t_b]$, such that the covariance matrix $\Sigma^o(t_b)$ is infimized for any arbitrarily fixed final time t_b. In other words, for any suboptimal observer gain matrix $P(.)$, the corresponding inferior error covariance matrix $\Sigma(t_b)$ will satisfy $\Sigma(t_b) - \Sigma^o(t_b) \geq 0$ (positive-semidefinite matrix). — This translates into the following

Statement of the optimal control problem:

Find an observer matrix $P : [t_a, t_b] \rightarrow R^{n \times p}$, such that the dynamic system

$$\dot{\Sigma}(t) = A(t)\Sigma(t) - P(t)C(t)\Sigma(t) + \Sigma(t)A^{\mathrm{T}}(t) - \Sigma(t)C^{\mathrm{T}}(t)P(t)^{\mathrm{T}}$$
$$+ B(t)Q(t)B^{\mathrm{T}}(t) + P(t)R(t)P^{\mathrm{T}}(t)$$

is transferred from the initial state

$$\Sigma(t_a) = \Sigma_a$$

to an unspecified final state $\Sigma(t_b)$ and such that the cost functional

$$J(P) = \Sigma(t_b) = \Sigma_a + \int_{t_a}^{t_b} \dot{\Sigma}(t) \, dt$$

is infimized.

The integrand in the cost functional is identical to the right-hand side of the differential equation of the state $\Sigma(t)$. Therefore, according to Chapter 2.2.3 and using the integral version of the cost functional, the correct formulation of the Hamiltonian is:

$$H = \lambda(t)\dot{\Sigma}(t)$$
$$= \lambda(t)\Big(A(t)\Sigma(t) - P(t)C(t)\Sigma(t) + B(t)Q(t)B^{\mathrm{T}}(t)$$
$$+ \Sigma(t)A^{\mathrm{T}}(t) - \Sigma(t)C^{\mathrm{T}}(t)P(t)^{\mathrm{T}} + P(t)R(t)P^{\mathrm{T}}(t)\Big)$$

with

$$\lambda(t_b) = I \in \mathcal{L}(\mathcal{X}_0, \mathcal{X}_0) \ ,$$

since the optimal control problem is regular.

Necessary conditions for superiority:

If $P^o : [t_0, t_b] \to R^{n \times p}$ is optimal, then the following conditions are satisfied:

a) Differential equations and boundary conditions:

$$\dot{\Sigma}^o = A\Sigma^o - P^oC\Sigma^o + \Sigma^oA^{\mathrm{T}} - \Sigma^oC^{\mathrm{T}}P^{o\mathrm{T}} + BQB^{\mathrm{T}} + P^oRP^{o\mathrm{T}}$$
$$\Sigma^o(t_a) = \Sigma_a$$
$$\dot{\lambda}^o = -\frac{\partial H}{\partial \Sigma}\Big|_o = -\lambda^oU(A - PC^o)$$
$$\lambda^o(t_b) = I \ .$$

b) Infimization of the Hamiltonian (see [3] or [12]):

$$\frac{\partial H}{\partial P}\Big|_o = \lambda^oU(P^oR - \Sigma^oC^{\mathrm{T}})T \equiv 0 \ .$$

Here, the following two operators have been used for ease of notation:

$$U : M \mapsto M + M^{\mathrm{T}} \qquad \text{for a quadratic matrix } M$$
$$T : N \mapsto N^{\mathrm{T}} \qquad \text{for an arbitrary matrix } N.$$

The infimization of the Hamiltonian yields the well-known optimal observer matrix

$$P^o(t) = \Sigma^o(t)C^{\mathrm{T}}(t)R^{-1}(t)$$

of the Kalman-Bucy Filter.

Plugging this result into the differential equation of the covariance matrix $\Sigma(t)$ leads to the following well-known matrix Riccati differential equation for the Kalman-Bucy Filter:

$$\dot{\Sigma}^o(t) = A(t)\Sigma^o(t) + \Sigma^o(t)A^{\mathrm{T}}(t)$$
$$- \Sigma^o(t)C^{\mathrm{T}}(t)R^{-1}(t)C(t)\Sigma^o(t) + B(t)Q(t)B^{\mathrm{T}}(t)$$
$$\Sigma^o(t_a) = \Sigma_a \ .$$

2.9 Exercises

1. Time-optimal damping of a harmonic oscillator:

   Find a piecewise continuous control $u : [0, t_b] \to [-1, +1]$, such that the dynamic system

   $$\begin{bmatrix} \dot{x}_1(t) \\ \dot{x}_2(t) \end{bmatrix} = \begin{bmatrix} 0 & 1 \\ -1 & 0 \end{bmatrix} \begin{bmatrix} x_1(t) \\ x_2(t) \end{bmatrix} + \begin{bmatrix} 0 \\ 1 \end{bmatrix} u(t)$$

   is transferred from the initial state

   $$\begin{bmatrix} x_1(0) \\ x_2(0) \end{bmatrix} = \begin{bmatrix} s_a \\ v_a \end{bmatrix}$$

   to the final state

   $$\begin{bmatrix} x_1(t_b) \\ x_2(t_b) \end{bmatrix} = \begin{bmatrix} 0 \\ 0 \end{bmatrix}$$

   in minimal time, i.e., such that the cost functional $J = \int_0^{t_b} dt$ is minimized.

2. Energy-optimal motion of an unstable system:

   Find an unconstrained optimal control $u : [0, t_b] \to R$, such that the dynamic system

   $$\dot{x}_1(t) = x_2(t)$$
   $$\dot{x}_2(t) = x_2 + u(t)$$

   is transferred from the initial state

   $$x_1(0) = 0$$
   $$x_2(0) = 0$$

   to a final state at the fixed final time t_b satisfying

   $$x_1(t_b) \geq s_b > 0$$
   $$x_2(t_b) \leq v_b$$

   and such that the cost functional

   $$J(u) = \int_0^{t_b} u^2(t)\, dt$$

   is minimized.

3. Fuel-optimal motion of a nonlinear system:

   Find a piecewise continuous control $u : [0, t_b] \to [0, +1]$, such that the dynamic system

   $$\dot{x}_1(t) = x_2(t)$$
   $$\dot{x}_2(t) = -x_2^2 + u(t)$$

is transferred from the given initial state

$$x_1(0) = 0$$
$$x_2(0) = v_a \qquad (0 < v_a < 1)$$

to the fixed final state at the fixed final time t_b

$$x_1(t_b) = s_b \qquad (s_b > 0)$$
$$x_2(t_b) = v_b \qquad (0 < v_b < 1)$$

and such that the cost functional

$$J(u) = \int_0^{t_b} u(t)\, dt$$

is minimized.

4. **LQ model-predictive control [2], [16]:**

Consider a linear dynamic system with the state vector $x(t) \in R^n$ and the unconstrained control vector $u(t) \in R^m$. All of the state variables are measured and available for state-feedback control. Some of the state variables are of particular interest. For convenience, they are collected in an output vector $y(t) \in R^p$ via the linear output equation

$$y(t) = C(t)x(t) .$$

Example: In a mechanical system, we are mostly interested in the state variables for the positions in all of the degrees of freedom, but much less in the associated velocities.

The LQ model-predictive tracking problem is formulated as follows:

Find $u : [t_a, t_b] \to R^m$ such that the linear dynamic system

$$\dot{x}(t) = A(t)x(t) + B(t)u(t)$$

is transferred from the given initial state $x(t_a) = x_a$ to an arbitrary final state $x(t_b)$ at the fixed final time t_b and such that the positive-definite cost functional

$$J(u) = \frac{1}{2}[y_d(t_b) - y(t_b)]^{\mathrm{T}} F_y [y_d(t_b) - y(t_b)]$$
$$+ \frac{1}{2} \int_{t_a}^{t_b} \Big([y_d(t) - y(t)]^{\mathrm{T}} Q_y(t)[y_d(t) - y(t)] + u^{\mathrm{T}}(t)R(t)u(t) \Big) dt$$

is minimized. The desired trajectory $y_d : [t_a, t_b] \to R^p$ is specified in advance. The weighting matrices F_y, $Q_y(t)$, and $R(t)$ are symmetric and positive-definite.

Prove that the optimal control law is the following combination of a feed-forward and a state feedback:

$$u(t) = R^{-1}(t)B^{\mathrm{T}}(t)w(t) - R^{-1}(t)B^{\mathrm{T}}(t)K(t)x(t)$$

where the n by n symmetric and positive-definite matrix $K(t)$ and the p-vector function $w(t)$ have to be calculated in advance for all $t \in [t_a, t_b]$ as follows:

$$\dot{K}(t) = -A^{\mathrm{T}}(t)K(t) - K(t)A(t)$$
$$\qquad\qquad + K(t)B(t)R^{-1}(t)B^{\mathrm{T}}(t)K(t) - C^{\mathrm{T}}(t)Q_y(t)C(t)$$
$$K(t_b) = C^{\mathrm{T}}(t_b)F_yC(t_b)$$
$$\dot{w}(t) = -[A(t) - B(t)R^{-1}(t)B^{\mathrm{T}}(t)K(t)]^{\mathrm{T}}w(t) - C(t)Q_y(t)y_d(t)$$
$$w(t_b) = C(t_b)F_y\, y_d(t_b)\ .$$

The resulting optimal control system is described by the following differential equation:

$$\dot{x}(t) = [A(t) - B(t)R^{-1}(t)B^{\mathrm{T}}(t)K(t)]x(t) + B(t)R^{-1}(t)B^{\mathrm{T}}(t)w(t)\ .$$

Note that $w(t)$ at any time t contains the information about the *future* of the desired output trajectory $y_d(.)$ over the remaining time interval $[t, t_b]$.

5. In Chapter 2.8.4, the Kalman-Bucy Filter has been derived. Prove that we have indeed infimized the Hamiltonian H. — We have only set the first derivative of the Hamiltonian to zero in order to find the known result.

3 Optimal State Feedback Control

Chapter 2 has shown how optimal control problems can be used by exploiting Pontryagin's Minimum Principle. Once the resulting two-point boundary value problem has been solved, the optimal control law is in an open-loop form: $u^o(t)$ for $t \in [t_a, t_b]$.

In principle, it is always possible to convert the optimal open-loop control law to an optimal closed-form control law by the following brute-force procedure: For every time $t \in [t_a, t_b]$, solve the "rest problem" of the original optimal control problem over the interval $[t, t_b]$ with the initial state $x(t)$. This yields the desired optimal control $u^o(x(t), t)$ at this time t which is a function of the present initial state $x(t)$. — Obviously, in Chapters 2.1.4, 2.1.5, and 2.3.4, we have found more elegant methods for converting the optimal open-loop control law into the corresponding optimal closed-loop control law.

The purpose of this chapter is to provide mathematical tools which allow us to find the optimal closed-loop control law directly. — Unfortunately, this leads to a partial differential equation for the "cost-to-go" function $\mathcal{J}(x, t)$ which needs to be solved.

3.1 The Principle of Optimality

Consider the following optimal control problem of Type B (see Chapter 2.4) with the fixed terminal time t_b:

Find an admissible control $u : [t_a, t_b] \rightarrow \Omega \subseteq R^m$, such that the constraints

$$x(t_a) = x_a$$
$$\dot{x}(t) = f(x(t), u(t), t) \qquad \text{for all } t \in [t_a, t_b]$$
$$x(t_b) \in S \subseteq R^n$$

are satisfied and such that the cost functional

$$J(u) = K(x(t_b)) + \int_{t_a}^{t_b} L(x(t), u(t), t) \, dt$$

is minimized.

Suppose that we have found the unique globally optimal solution with the optimal control trajectory $u^o : [t_a, t_b] \to \Omega \subseteq R^m$ and the corresponding optimal state trajectory $x^o : [t_a, t_b] \to R^n$ which satisfies $x^o(t_a) = x_a$ and $x^o(t_b) \in S$.

Now, pick an arbitrary time $\tau \in (t_a, t_b)$ and bisect the original optimal control problem into an antecedent optimal control problem over the time interval $[t_a, \tau]$ and a succedent optimal problem over the interval $[\tau, t_b]$.

The *antecedent* optimal control problem is:

Find an admissible control $u : [t_a, \tau] \to \Omega$, such that the dynamic system
$$\dot{x}(t) = f(x(t), u(t), t)$$

is transferred from the initial state
$$x(t_a) = x_a$$

to the fixed final state
$$x(\tau) = x^o(\tau)$$

at the fixed final time τ and such that the cost functional
$$J(u) = \int_{t_a}^{\tau} L(x(t), u(t), t)\, dt$$
is minimized.

The *succedent* optimal control problem is:

Find an admissible control $u : [\tau, t_b] \to \Omega$, such that the dynamic system
$$\dot{x}(t) = f(x(t), u(t), t)$$

is transferred from the given initial state
$$x(\tau) = x^o(\tau)$$

to the partially constrained final state
$$x(t_b) \in S$$

at the fixed final time t_b and such that the cost functional
$$J(u) = K(x(t_b)) + \int_{\tau}^{t_b} L(x(t), u(t), t)\, dt$$
is minimized.

The following important but almost trivial facts can easily be derived:

Theorem: The Principle of Optimality

1) The optimal solution of the succedent optimal control problem coincides with the succedent part of the optimal solution of the original problem.

2) The optimal solution of the antecedent optimal control problem coincides with the antecedent part of the optimal solution of the original problem.

Note that only the first part is relevant to the method of dynamic programming and to the Hamilton-Jacobi-Bellman Theory (Chapter 3.2).

Proof

1) Otherwise, combining the optimal solution of the succedent optimal control problem with the antecedent part of the solution of the original optimal control problem would yield a better solution of the latter.

2) Otherwise, combining the optimal solution of the antecedent optimal control problem with the succedent part of the solution of the original optimal control problem would yield a better solution of the latter.

Conceptually, we can solve the succedent optimal control problem for any arbitrary initial state $x \in R^n$ at the initial time τ, rather than for the fixed value $x^o(\tau)$ only. Furthermore, we can repeat this process for an arbitrary initial time $t \in [t_a, t_b]$, rather than for the originally chosen value τ only. Concentrating only on the optimal value of the cost functional in all of these cases yields the so-called *optimal cost-to-go function*

$$\mathcal{J}(x,t) = \min_{u(\cdot)} \left\{ K(x(t_b)) + \int_t^{t_b} L(x(t), u(t), t)\, dt \ \Big|\ x(t) = x \right\} .$$

Working with the optimal cost-to-go function, the Principle of Optimality reveals two additional important but almost trivial facts:

Lemma

3) The optimal solution of an antecedent optimal control problem with a *free* final state at the fixed final time τ and with the cost functional

$$J = \mathcal{J}(x(\tau), \tau) + \int_{t_a}^{\tau} L(x(t), u(t), t)\, dt$$

coincides with the antecedent part of the optimal solution of the original optimal control problem.

4) The optimal costate vector $\lambda^o(\tau)$ corresponds to the gradient of the optimal cost-to-go function, i.e.,

$$\lambda^o(\tau) = \nabla_x \mathcal{J}(x^o(\tau), \tau) \quad \text{for all } \tau \in [t_a, t_b] ,$$

provided that $\mathcal{J}(x, \tau)$ is continuously differentiable with respect to x at $x^o(\tau)$.

Proof

3) Otherwise, combining the optimal solution of the modified antecedent optimal control problem with the succedent part of the solution of the original optimal control problem would yield a better solution of the latter.

4) This is the necessary condition of Pontryagin's Minimum Principle for the final costate in an optimal control problem with a free final state, where the cost functional includes a final state penalty term (see Chapter 2.3.2, Theorem C).

3.2 Hamilton-Jacobi-Bellman Theory

3.2.1 Sufficient Conditions for the Optimality of a Solution

Consider the usual formulation of an optimal control problem with an unspecified final state at the fixed final time:

Find a piecewise continuous control $u : [t_a, t_b] \rightarrow \Omega$ such that the dynamic system

$$\dot{x}(t) = f(x(t), u(t), t)$$

is transferred from the given initial state $x(t_a) = x_a$ to an arbitrary final state at the fixed final time t_b and such that the cost functional

$$J(u) = K(x(t_b)) + \int_{t_a}^{t_b} L(x(t), u(t), t)\, dt$$

is minimized.

Since the optimal control problem is regular with $\lambda_0^o = 1$, the Hamiltonian function is

$$H(x, u, \lambda, t) = L(x, u, t) + \lambda^\mathrm{T} f(x, u, t)\ .$$

Let us introduce the $n+1$-dimensional set $Z = X \times [a, b] \subseteq R^n \times R$, where X is a (hopefully very large) subset of the state space R^n with non-empty interior and $[a, b]$ is a subset of the time axis containing at least the interval $[t_a, t_b]$, as shown in Fig. 3.1.

Let us consider arbitrary admissible controls $\widehat{u} : [t_a, t_b] \rightarrow \Omega$ which generate the corresponding state trajectories $\widehat{x} : [t_a, t_b] \rightarrow R^n$ starting at $x(t_a) = x_a$. We are mainly interested in state trajectories which do not leave the set Z, i.e., which satisfy $x(t) \in X$ for all $t \in [t_a, t_b]$.

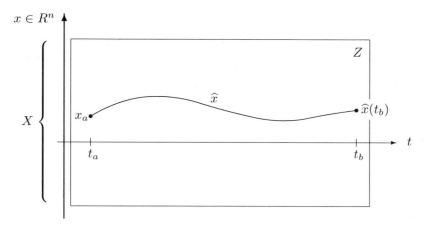

Fig. 3.1. Example of a state trajectory $\widehat{x}(.)$ which does not leave X.

With the following hypotheses, the sufficient conditions for the global optimality of a solution of an optimal control problem can be stated in the Hamilton-Bellman-Jacobi Theorem below.

Hypotheses

a) Let $\widehat{u} : [t_a, t_b] \to \Omega$ be an admissible control generating the state trajectory $\widehat{x} : [t_a, t_b] \to R^n$ with $\widehat{x}(t_a) = x_a$ and $\widehat{x}(.) \in Z$.

b) For all $(x,t) \in Z$ and all $\lambda \in R^n$, let the Hamiltonian function $H(x, \omega, \lambda, t) = L(x, \omega, t) + \lambda^T f(x, \omega, t)$ have a unique global minimum with respect to $\omega \in \Omega$ at

$$\omega = \widetilde{u}(x, \lambda, t) \in \Omega .$$

c) Let $\mathcal{J}(x,t) : Z \to R$ be a continuously differentiable function satisfying the Hamilton-Jacobi-Bellman partial differential equation

$$\frac{\partial \mathcal{J}(x,t)}{\partial t} + H\left[x, \widetilde{u}\left(x, \nabla_x \mathcal{J}(x,t), t\right), \nabla_x \mathcal{J}(x,t), t \right] = 0$$

with the boundary condition

$$\mathcal{J}(x, t_b) = K(x) \quad \text{for all } (x, t_b) \in Z .$$

Remarks:

- The function $\widetilde{u}$ is called the H-minimizing control.
- When hypothesis b is satisfied, the Hamiltonian H is said to be "normal".

Hamilton-Jacobi-Bellman Theorem

If the hypotheses a, b, and c are satisfied and if the control trajectory $\widehat{u}(.)$ and the state trajectory $\widehat{x}(.)$ which is generated by $\widehat{u}(.)$ are related via

$$\widehat{u}(t) = \widetilde{u}\Big(\widehat{x}(t), \nabla_x \mathcal{J}(\widehat{x}(t), t), t\Big) ,$$

then the solution $\widehat{u}, \widehat{x}$ is optimal with respect to all state trajectories x generated by an admissible control trajectory u, which do not leave X. Furthermore, $\mathcal{J}(x, t)$ is the optimal cost-to-go function.

Lemma

If $Z = R^n \times [t_a, t_b]$, then the solution $\widehat{u}, \widehat{x}$ is globally optimal.

Proof

For a complete proof of these sufficiency conditions see [2, pp. 351–363].

3.2.2 Plausibility Arguments about the HJB Theory

In this section, a brief reasoning is given as to why the Hamilton-Jacobi-Bellman partial differential equation pops up.

We have the following facts:

1) If the Hamiltonian function H is normal, we have the following unique H-minimizing optimal control:

$$u^o(t) = \widetilde{u}\Big(x^o(t), \lambda^o(t), t\Big) .$$

2) The optimal cost-to-go function $\mathcal{J}(x, t)$ must obviously satisfy the boundary condition

$$\mathcal{J}(x, t_b) = K(x)$$

because at the final time t_b, the cost functional only consists of the final state penalty term $K(x)$.

3) The Principle of Optimality has shown that the optimal costate $\lambda^o(t)$ corresponds to the gradient of the optimal cost-to-go function,

$$\lambda^o(t) = \nabla_x \mathcal{J}(x^o(t), t) ,$$

wherever $\mathcal{J}(x^o(t), t)$ is continuously differentiable with respect to x at $x = x^o(t)$.

4) Along an arbitrary admissible trajectory $u(.)$, $x(.)$, the corresponding suboptimal cost-to-go function

$$J(x(t), t) = K(x(t_b)) + \int_t^{t_b} L(x(t), u(t), t)\, dt$$

evolves according to the following differential equation:

$$\frac{dJ}{dt} = \frac{\partial J}{\partial x} \dot{x} + \frac{\partial J}{\partial t} = \lambda^{\mathrm{T}} f(x, u, t) + \frac{\partial J}{\partial t} = -L(x, u, t) \ .$$

Hence,

$$\frac{\partial J}{\partial t} = -\lambda^{\mathrm{T}} f(x, u, t) - L(x, u, t) = -H(x, u, \lambda, t) \ .$$

This corresponds to the partial differential equation for the optimal cost-to-go function $\mathcal{J}(x, t)$, except that the optimal control law has not been plugged in yet.

3.2.3 The LQ Regulator Problem

A simpler version of the LQ regulator problem considered here has been stated in Problem 5 (Chapter 1, p. 8) and analyzed in Chapter 2.3.4.

Statement of the optimal control problem

Find an optimal state feedback control law $u : R^n \times [t_a, t_b] \to R^m$, such that the linear dynamic system

$$\dot{x}(t) = A(t)x(t) + B(t)u(t)$$

is transferred from the given initial state $x(t_a) = x_a$ to an arbitrary final state at the fixed final time t_b and such that the quadratic cost functional

$$J(u) = \frac{1}{2} x^{\mathrm{T}}(t_b) F x(t_b)$$
$$+ \int_{t_a}^{t_b} \left(\frac{1}{2} x^{\mathrm{T}}(t) Q(t) x(t) + x^{\mathrm{T}}(t) N(t) u(t) + \frac{1}{2} u^{\mathrm{T}}(t) R(t) u(t) \right) dt$$

is minimized, where $R(t)$ is symmetric and positive-definite, and F, Q, and $\begin{bmatrix} Q(t) & N(t) \\ N^{\mathrm{T}}(t) & R(t) \end{bmatrix}$ are symmetric and positive-semidefinite.

Analysis of the problem

The Hamiltonian function

$$H = \frac{1}{2} x^{\mathrm{T}} Q x + x^{\mathrm{T}} N u + \frac{1}{2} u^{\mathrm{T}} R u + \lambda^{\mathrm{T}} A x + \lambda^{\mathrm{T}} B u$$

has the following H-minimizing control:

$$u = -R^{-1}[B^{\mathrm{T}}\lambda + N^{\mathrm{T}}x] = -R^{-1}[B^{\mathrm{T}}\nabla_x \mathcal{J} + N^{\mathrm{T}}x] \ .$$

Thus, the resulting Hamilton-Jacobi-Bellman partial differential equation is

$$
\begin{aligned}
0 &= \frac{\partial \mathcal{J}}{\partial t} + H(x, \tilde{u}(x, \nabla_x \mathcal{J}, t), \nabla_x \mathcal{J}, t) \\
&= \frac{\partial \mathcal{J}}{\partial t} + \frac{1}{2}\Big(x^{\mathrm{T}}Qx - x^{\mathrm{T}}NR^{-1}N^{\mathrm{T}}x - \nabla_x \mathcal{J}^{\mathrm{T}}BR^{-1}B^{\mathrm{T}}\nabla_x \mathcal{J} \\
&\qquad + \nabla_x \mathcal{J}^{\mathrm{T}}[A - BR^{-1}N^{\mathrm{T}}]x + x^{\mathrm{T}}[A - BR^{-1}N^{\mathrm{T}}]^{\mathrm{T}}\nabla_x \mathcal{J} \Big)
\end{aligned}
$$

with the boundary condition

$$\mathcal{J}(x, t_b) = \frac{1}{2}x^{\mathrm{T}}Fx \ .$$

Obviously, an ansatz for the optimal cost-to-go function $\mathcal{J}(x, t)$ which is quadratic in x should work. This results in a partial differential equation in which all of the terms are quadratic in x. The ansatz

$$\mathcal{J}(x, t) = \frac{1}{2}x^{\mathrm{T}}K(t)x$$

leads to

$$\nabla_x \mathcal{J}(x, t) = K(t)x \quad \text{and} \quad \frac{\partial \mathcal{J}(x, t)}{\partial t} = \frac{1}{2}x^{\mathrm{T}}\dot{K}(t)x \ .$$

The following final form of the Hamilton-Jacobi-Bellman partial differential equation is obtained:

$$
\begin{aligned}
\frac{1}{2}x^{\mathrm{T}}\Big(\dot{K}(t) + Q - NR^{-1}N^{\mathrm{T}} - K(t)BR^{-1}B^{\mathrm{T}}K(t) \\
+ K(t)[A - BR^{-1}N^{\mathrm{T}}] + [A - BR^{-1}N^{\mathrm{T}}]^{\mathrm{T}}K(t) \Big)x = 0
\end{aligned}
$$

$$\mathcal{J}(x, t_b) = \frac{1}{2}x^{\mathrm{T}}Fx \ .$$

Therefore, we get the following optimal state feedback control law:

$$u(t) = -R^{-1}(t)[B^{\mathrm{T}}(t)K(t) + N^{\mathrm{T}}(t)]x(t) \ ,$$

where the symmetric and positive-(semi)definite matrix $K(t)$ has to be computed in advance by solving the matrix Riccati differential equation

$$
\begin{aligned}
\dot{K}(t) = {}& -[A(t) - B(t)R^{-1}(t)N^{\mathrm{T}}(t)]^{\mathrm{T}}K(t) - K(t)[A(t) - B(t)R^{-1}(t)N^{\mathrm{T}}(t)] \\
& - K(t)B(t)R^{-1}(t)B^{\mathrm{T}}(t)K(t) - Q(t) + N(t)R^{-1}(t)N^{\mathrm{T}}(t)
\end{aligned}
$$

with the boundary condition

$$K(t_b) = F \ .$$

3.2.4 The Time-Invariant Case with Infinite Horizon

In this section, time-invariant optimal control problems with the unconstrained control vector $u(t) \in R^m$, an infinite horizon, and a free final state
$x(t_b)$ at the infinite final time $t_b = \infty$ are considered.

The most general statement of this optimal control problem is:

Find a piecewise continuous control $u : [0, \infty) \to R^m$, such that the completely controllable dynamic system

$$\dot{x}(t) = f(x(t), u(t))$$

is transferred from the given initial state

$$x(0) = x_a$$

to an arbitrary final state $x(\infty) \in R^n$ at the infinite final time and such that
the cost functional

$$J(u) = \int_0^\infty L(x(t), u(t)) \, dt$$

is minimized and attains a finite optimal value.

In order to have a well-posed problem, the variables of the problem should be
chosen in such a way that the intended stationary equilibrium state is at $x = 0$
and that it can be reached by an asymptotically vanishing control $u(t) \to 0$
as $t \to \infty$. Therefore, $f(0,0) = 0$ is required. — Furthermore, choose the
integrand L of the cost functional with $L(0,0) = 0$ and such that it is strictly
convex in both x and u and such that $L(x, u)$ grows without bound, for all
(x, u) where either x, or u, or both x and u go to infinity in any direction in
the state space R^n or the control space R^m, respectively. — Of course, we
assume that both f and L are at least once continuously differentiable with
respect to x and u.

Obviously, in the time-invariant case with infinite horizon, the optimal cost-
to-go function $\mathcal{J}(x, t)$ is time-invariant, i.e.,

$$\mathcal{J}(x, t) \equiv \mathcal{J}(x) \ ,$$

because the optimal solution is shift-invariant. It does not matter whether
the system starts with the initial state x_a at the initial time 0 or with the
same initial state x_a at some other initial time $t_a \neq 0$.

Therefore, the Hamilton-Jacobi-Bellman partial differential equation

$$\frac{\partial \mathcal{J}(x, t)}{\partial t} + H\left[x, \tilde{u}\left(x, \nabla_x \mathcal{J}(x, t), t \right), \nabla_x \mathcal{J}(x, t), t \right] = 0$$

degenerates to the partial differential equation

$$H\left[x, \tilde{u}\left(x, \nabla_x \mathcal{J}(x)\right), \nabla_x \mathcal{J}(x)\right] = 0$$

and loses the former boundary condition $\mathcal{J}(x, t_b) = K(x)$.

In the special case of a dynamic system of first order ($n = 1$), this is an ordinary differential equation which can be integrated using the boundary condition $\mathcal{J}(0) = 0$.

For dynamic systems of higher order ($n \geq 2$), the following alternative problem solving techniques are available:

a) Choose an arbitrary positive-definite function $K(x)$ with $K(0) = 0$ which satisfies the usual growth condition. Integrate the Hamilton-Jacobi-Bellman partial differential equation over the region $R^n \times (-\infty, t_b]$ using the boundary condition $\mathcal{J}(x, t_b) = K(x)$. The solution $\mathcal{J}(x, t)$ asymptotically converges to the desired time-invariant optimal cost-to-go function $\mathcal{J}(x)$ as $t \to -\infty$,

$$\mathcal{J}(x) = \lim_{t \to -\infty} \mathcal{J}(x, t) .$$

b) Solve the two equations

$$H(x, u, \lambda) = 0$$
$$\nabla_u H(x, u, \lambda) = 0$$

in order to find the desired optimal state feedback control law $u^o(x)$ without calculating the optimal cost-to-go function $\mathcal{J}(x)$. Since both of these equations are linear in the costate λ, there is a good chance[3] that λ can be eliminated without calculating $\lambda = \nabla_x \mathcal{J}(x)$ explicitly. This results in an implicit form of the optimal state feedback control law.

Example

Let us assume that we have already solved the following LQ regulator problem for an unstable dynamic system of first order:

$$\dot{x}(t) = ax(t) + bu(t) \quad \text{with } a > 0, \ b \neq 0$$

$$x(0) = x_a$$

$$J(u) = \frac{1}{2} \int_0^\infty \left(qx^2(t) + u^2(t)\right) dt \quad \text{with } q > 0 .$$

---

[3] at least in the case $n=1$ and hopefully also in the case $n=m+1$

The result is the linear state feedback controller

$$u(t) = -gx(t)$$

with

$$g = bk = \frac{a + a\sqrt{1 + \frac{b^2 q}{a^2}}}{b} .$$

Now, we want to replace this linear controller by a nonlinear one which is "softer" for large values of $|x|$, i.e., which shows a suitable saturation behavior, but which retains the "stiffness" of the LQ regulator for small signals x. — Note that, due to the instability of the plant, the controller must not saturate to a constant maximal value for the control. Rather, it can only saturate to a "softer" linear controller of the form $u = -g_\infty x$ for large $|x|$ with $g_\infty > a/b$.

In order to achieve this goal, the cost functional is modified as follows:

$$J(u) = \frac{1}{2} \int_0^\infty \left(qx^2(t) + u^2(t) + \beta u^4(t) \right) dt \quad \text{with } \beta > 0 .$$

According to the work-around procedure b, the following two equations must be solved:

$$H(x, u, \lambda) = \frac{q}{2}x^2 + \frac{1}{2}u^2 + \frac{\beta}{2}u^4 + \lambda ax + \lambda bu = 0$$

$$\frac{\partial H}{\partial u} = u + 2\beta u^3 + \lambda b = 0 .$$

Eliminating λ yields the implicit optimal state feedback control law

$$3\beta u^4 + \frac{4\beta a}{b}xu^3 + u^2 + \frac{2a}{b}xu - qx^2 = 0 .$$

The explicit optimal state feedback control is obtained by solving this equation for the unique stabilizing controller $u(x)$:

$$u(x) = \arg \left\{ 3\beta u^4 + \frac{4\beta a}{b}xu^3 + u^2 + \frac{2a}{b}xu - qx^2 = 0 \; \middle| \; \begin{array}{ll} u < -\frac{a}{b}x & \text{for } x > 0 \\ u = 0 & \text{for } x = 0 \\ u > -\frac{a}{b}x & \text{for } x < 0 \end{array} \right\} .$$

The small-signal characteristic is identical with the characteristic of the LQ regulator because the fourth-order terms u^4 and xu^3 are negligible. Conversely, for the large-signal characteristic, the fourth-order terms dominate and the second-order terms u^2, xu, and x^2 are negligible. Therefore, the large-signal characteristic is

$$u \approx -\frac{4a}{3b}x .$$

In Fig. 3.2, the nonlinear optimal control law is depicted for the example where $a = 1$, $b = 1$, $q = 8$, and $\beta = 1$ with the LQ regulator gain $g = 4$ and the large-signal gain $g_\infty = -\frac{4}{3}$.

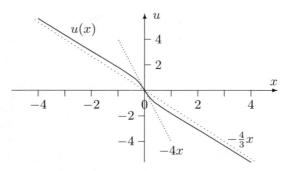

Fig. 3.2. Optimal nonlinear control law.

The inclined reader is invited to verify that replacing the term βu^4 in the cost functional by βu^{2k} with $k \geq 2$ results in the large-gain characteristic

$$u \approx -\frac{2ka}{(2k-1)b}\,x \ .$$

3.3 Approximatively Optimal Control

In most cases, no analytical solution of the Hamilton-Jacobi-Bellman partial differential equation can be found. Furthermore, solving it numerically can be extremely cumbersome.

Therefore, a method is presented here which allows to find approximate solutions for the state feedback control law for a time-invariant optimal control problem with an infinite horizon.

This method has been proposed by Lukes in [27]. It is well suited for problems where the right-hand side $f(x, u)$ of the state differential equation, and the integrand $L(x, u)$ of the cost functional, and the optimal state feedback control law $u(x)$ can be expressed by polynomial approximations around the equilibrium $x \equiv 0$ and $u \equiv 0$, which converge rapidly. (Unfortunately, the problem presented in Chapter 3.2.4 does not belong to this class.)

Let us consider a time-invariant optimal control problem with an infinite horizon for a nonlinear dynamic system with a non-quadratic cost functional, which is structured as follows:

Find a time-invariant optimal state feedback control law $u : R^n \to R^m$, such that the dynamic system

$$\dot{x}(t) = F(x(t), u(t)) = Ax(t) + Bu(t) + f(x(t), u(t))$$

is transferred from an arbitrary initial state $x(0) = x_a$ to the equilibrium state

$x = 0$ at the infinite final time and such that the cost functional

$$J(u) = \int_0^\infty L(x(t), u(t))\, dt$$

$$= \int_0^\infty \left(\frac{1}{2} x^{\mathrm{T}}(t) Q x(t) + x^{\mathrm{T}}(t) N u(t) + \frac{1}{2} u^{\mathrm{T}}(t) R u(t) + \ell(x(t), u(t)) \right) dt$$

is minimized.

In this problem statement, it is assumed that the following conditions are satisfied:

- $[A, B]$ stabilizable
- $R > 0$
- $Q = C^{\mathrm{T}} C \geq 0$
- $[A, C]$ detectable
- $\begin{bmatrix} Q & N \\ N^{\mathrm{T}} & R \end{bmatrix} \geq 0$
- $f(x, u)$ contains only second-order or higher-order terms in x and/or u
- $\ell(x, u)$ contains only third-order or higher-order terms in x and/or u.

3.3.1 Notation

Here, some notation is introduced for derivatives and for sorting the terms of the same order in a polynomial approximation of a scalar-valued or a vector-valued function around a reference point.

Differentiation

For the Jacobian matrix of the partial derivatives of an n-vector-valued function f with respect to the m-vector u, the following symbol is used:

$$f_u = \frac{\partial f}{\partial u} = \begin{bmatrix} \dfrac{\partial f_1}{\partial u_1} & \cdots & \dfrac{\partial f_1}{\partial u_m} \\ \vdots & & \vdots \\ \dfrac{\partial f_n}{\partial u_1} & \cdots & \dfrac{\partial f_n}{\partial u_m} \end{bmatrix}.$$

Note that a row vector results for the derivative of a scalar function f.

Sorting Powers

In polynomial functions, we collect the terms of the same power as follows:

$$f(x, u) = f^{(2)}(x, u) + f^{(3)}(x, u) + f^{(4)}(x, u) + \ldots$$
$$\ell(x, u) = \ell^{(3)}(x, u) + \ell^{(4)}(x, u) + \ell^{(5)}(x, u) + \ldots$$
$$\mathcal{J}(x) = \mathcal{J}^{(2)}(x) + \mathcal{J}^{(3)}(x) + \mathcal{J}^{(4)}(x) + \ldots$$
$$u^o(x) = u^{o(1)}(x) + u^{o(2)}(x) + u^{o(3)}(x) + \ldots \quad .$$

Example: In the simplest case of a scalar function ℓ with the scalar arguments x and u, the function $\ell^{(3)}(x, u)$ has the following general form:

$$\ell^{(3)}(x, u) = \alpha x^3 + \beta x^2 u + \gamma x u^2 + \delta u^3 \ .$$

For the derivatives of the functions f and ℓ, the powers are sorted in the analogous way, e.g.,

$$f_u(x, u) = f_u^{(1)}(x, u) + f_u^{(2)}(x, u) + f_u^{(3)}(x, u) + \dots$$
$$\ell_u(x, u) = \ell_u^{(2)}(x, u) + \ell_u^{(3)}(x, u) + \ell_u^{(4)}(x, u) + \dots \ \ .$$

Notice the following fact for the derivative of a function with respect to a scalar-valued or a vector-valued argument:

$$\ell_u^{(k)}(x, u) = \frac{\partial \ell^{(k+1)}(x, u)}{\partial u} \ .$$

In the previous example, we get

$$\ell_u^{(2)}(x, u) = \frac{\partial \ell^{(3)}(x, u)}{\partial u} = \beta x^2 + 2\gamma x u + 3\delta u^2 \ .$$

In general, this kind of notation will be used in the sequel. There is one exception though: In order to have the notation for the derivatives of the cost-to-go function $\mathcal{J}$ match the notation used by Lukes in [27], we write

$$\mathcal{J}_x(x) = \mathcal{J}_x^{[2]}(x) + \mathcal{J}_x^{[3]}(x) + \mathcal{J}_x^{[4]}(x) + \dots$$

instead of $\mathcal{J}_x^{(1)}(x) + \mathcal{J}_x^{(2)}(x) + \mathcal{J}_x^{(3)}(x) \dots \ \ .$ — Note the difference in the "exponents" and their brackets.

3.3.2 Lukes' Method

In the first approximation step, the linear controller $u^{o(1)}$ is determined by solving the LQ regulator problem for the linearized dynamic system and for the purely quadratic part of the cost functional.

In each additional approximation step of Lukes' recursive method, one additional power is added to the feedback law $u(x)$, while one additional power of the approximations of f and ℓ is processed.

As shown in Chapter 3.2.4, the following two equations have to be solved approximately in each approximation step:

$$H = \mathcal{J}_x(x)F(x, u) \ + L(x, u) \ = 0 \tag{1}$$

$$H_u = \mathcal{J}_x(x)F_u(x, u) + L_u(x, u) = 0 \ . \tag{2}$$

In the problem at hand, we have the following equations:

$$H = \mathcal{J}_x(x)[Ax+Bu+f(x,u)]$$
$$+ \frac{1}{2}x^\mathrm{T}Qx + x^\mathrm{T}Nu + \frac{1}{2}u^\mathrm{T}Ru + \ell(x,u) = 0 \tag{3}$$
$$H_u = \mathcal{J}_x(x)(B+f_u(x,u)) + x^\mathrm{T}N + u^\mathrm{T}R + \ell_u(x,u) = 0 \ . \tag{4}$$

Solving the implicit equation (4) for u^o yields:

$$u^{o\mathrm{T}} = -[\mathcal{J}_x(x)(B+f_u(x,u^o)) + x^\mathrm{T}N + \ell_u(x,u^o)]R^{-1} \ . \tag{4'}$$

1$^\mathrm{st}$ Approximation: LQ-Regulator

$$\dot{x}(t) = Ax + Bu$$
$$J(u) = \int_0^\infty \left(\frac{1}{2}x^\mathrm{T}Qx + x^\mathrm{T}Nu + \frac{1}{2}u^\mathrm{T}Ru\right) dt$$
$$u^{o(1)} = Gx \quad \text{with} \quad G = -R^{-1}(B^\mathrm{T}K + N^\mathrm{T}) \ ,$$

where K is the unique stabilizing solution of the matrix Riccati equation

$$[A-BR^{-1}N^\mathrm{T}]^\mathrm{T}K+K[A-BR^{-1}N^\mathrm{T}]-KBR^{-1}B^\mathrm{T}K+Q-NR^{-1}N^\mathrm{T}=0 \ .$$

The resulting linear control system is described by the differential equation

$$\dot{x}(t) = [A+BG]x(t) = A^o x(t)$$

and has the cost-to-go function

$$\mathcal{J}^{(2)}(x) = \frac{1}{2}x^\mathrm{T}Kx \quad \text{with} \quad \mathcal{J}_x^{[2]}(x) = x^\mathrm{T}K \ .$$

2$^\mathrm{nd}$ Approximation

$$u^o(x) = u^{o(1)}(x) + u^{o(2)}(x)$$
$$\mathcal{J}_x(x) = \mathcal{J}_x^{[2]}(x) + \mathcal{J}_x^{[3]}(x)$$

a) Determining $\mathcal{J}_x^{[3]}(x)$:

Using (3) yields:

$$0 = (\mathcal{J}_x^{[2]} + \mathcal{J}_x^{[3]})[Ax + B(u^{o(1)}+u^{o(2)}) + f(x, u^{o(1)}+u^{o(2)})]$$
$$+ \frac{1}{2}x^\mathrm{T}Qx + x^\mathrm{T}N(u^{o(1)}+u^{o(2)}) + \frac{1}{2}(u^{o(1)}+u^{o(2)})^\mathrm{T}R(u^{o(1)}+u^{o(2)})$$
$$+ \ell(x, u^{o(1)}+u^{o(2)})$$

Cubic terms:

$$0 = \mathcal{J}_x^{[3]}[Ax + Bu^{o(1)}] + \mathcal{J}_x^{[2]}[Bu^{o(2)} + f^{(2)}(x, u^{o(1)})]$$
$$+ x^T N u^{o(2)} + \frac{1}{2} u^{o(1)T} R u^{o(2)} + \frac{1}{2} u^{o(2)T} R u^{o(1)} + \ell^{(3)}(x, u^{o(1)})$$
$$= \mathcal{J}_x^{[3]} A^o x + \mathcal{J}_x^{[2]} f^{(2)}(x, u^{o(1)}) + \ell^{(3)}(x, u^{o(1)})$$
$$+ \underbrace{[\mathcal{J}_x^{[2]} B + x^T N + u^{o(1)T} R]}_{=0} u^{o(2)}$$

Therefore, the equation for $\mathcal{J}_x^{[3]}(x)$ is:

$$0 = \mathcal{J}_x^{[3]} A^o x + \mathcal{J}_x^{[2]} f^{(2)}(x, u^{o(1)}) + \ell^{(3)}(x, u^{o(1)}) \quad . \tag{6}$$

b) Determining $u^{o(2)}(x)$:

Using (4′) yields:

$$(u^{o(1)} + u^{o(2)})^T = -\Big[(\mathcal{J}_x^{[2]} + \mathcal{J}_x^{[3]})(B + f_u(x, u^{o(1)} + u^{o(2)})$$
$$+ x^T N + \ell_u(x, u^{o(1)} + u^{o(2)})\Big] R^{-1}$$

Quadratic terms:

$$u^{o(2)T} = -\Big[\mathcal{J}_x^{[3]} B + \mathcal{J}_x^{[2]} f_u^{(1)}(x, u^{o(1)}) + \ell_u^{(2)}(x, u^{o(1)})\Big] R^{-1} \tag{7}$$

Note that in the equations (6) and (7), $u^{o(2)}$ does not contribute to the right-hand sides. Therefore, these two equations are decoupled. Equation (7) is an explicit equation determining $u^{o(2)}$. — This feature appears in an analogous way in all of the successive approximation steps.

3rd Approximation:

$$u^*(x) = u^{o(1)}(x) + u^{o(2)}(x)$$
$$u^o(x) = u^*(x) + u^{o(3)}(x)$$
$$\mathcal{J}_x(x) = \mathcal{J}_x^{[2]}(x) + \mathcal{J}_x^{[3]}(x) + \mathcal{J}_x^{[4]}(x)$$

a) Determining $\mathcal{J}_x^{[4]}(x)$:

$$0 = \mathcal{J}_x^{[4]} A^o x + \mathcal{J}_x^{[3]} B u^{o(2)}$$
$$+ \mathcal{J}_x^{[3]} f^{(2)}(x, u^*) + \mathcal{J}_x^{[2]} f^{(3)}(x, u^*)$$
$$+ \frac{1}{2} u^{o(2)T} R u^{o(2)} + \ell^{(4)}(x, u^*)$$

b) Determining $u^{o(3)}(x)$:

$$u^{o(3)\,\mathrm{T}} = -[\mathcal{J}_x^{[4]}B + \mathcal{J}_x^{[3]}f_u^{(1)}(x,u^*) + \mathcal{J}_x^{[2]}f_u^{(2)}(x,u^*) + \ell_u^{(3)}(x,u^*)]R^{-1}$$

k^{th} Approximation $(k \geq 4)$

$$u^*(x) = \sum_{i=1}^{k-1} u^{o(i)}$$

$$u^o(x) = u^*(x) + u^{o(k)}(x)$$

$$\mathcal{J}_x(x) = \sum_{j=2}^{k+1} \mathcal{J}_x^{[j]}(x)$$

a) Determining $\mathcal{J}_x^{[k+1]}(x)$:

For k even:

$$0 = \mathcal{J}_x^{[k+1]}A^o x + \sum_{j=2}^{k-1} \mathcal{J}_x^{[k+2-j]}Bu^{o(j)} + \sum_{j=2}^{k} \mathcal{J}_x^{[k+2-j]}f^{(j)}(x,u^*)$$

$$+ \sum_{j=2}^{\frac{k}{2}} u^{o(j)\,\mathrm{T}}Ru^{o(k+1-j)} + \ell^{(k+1)}(x,u^*)$$

For k odd:

$$0 = \mathcal{J}_x^{[k+1]}A^o x + \sum_{j=2}^{k-1} \mathcal{J}_x^{[k+2-j]}Bu^{o(j)} + \sum_{j=2}^{k} \mathcal{J}_x^{[k+2-j]}f^{(j)}(x,u^*)$$

$$+ \sum_{j=2}^{\frac{k-1}{2}} u^{o(j)\,\mathrm{T}}Ru^{o(k+1-j)} + \frac{1}{2}u^{o(\frac{k+1}{2})\,\mathrm{T}}Ru^{o(\frac{k+1}{2})} + \ell^{(k+1)}(x,u^*)$$

b) Determining $u^{o(k)}(x)$:

$$u^{o(k)\,\mathrm{T}} = -\left[\mathcal{J}_x^{[k+1]}B + \ell_u^{(k)}(x,u^*) + \sum_{j=1}^{k-1} \mathcal{J}_x^{[k+1-j]}f_u^{(j)}(x,u^*)\right]R^{-1}$$

These formulae are valid for $k \geq 2$ already, if the value of a void sum is defined to be zero.

3.3.3 Controller with a Progressive Characteristic

For a linear time-invariant dynamic system of first order, we want to design a time-invariant state feedback control $u(x)$, the characteristic of which is super-linear, i.e., $u(x)$ is progressive for larger values of the state x.

In order to achieve this goal, we formulate a cost functional which penalizes the control quadratically and the state super-quadratically.

As an example, let us consider the optimal state feedback control problem described by the following equations:

$$\dot{x}(t) = ax(t) + u(t)$$

$$J(u) = \int_0^\infty \left(q\cosh(x(t)) - q + \frac{1}{2}u^2(t) \right) dt \ ,$$

where a and q are positive constants.

Using the series expansion

$$\cosh(x) = 1 + \frac{x^2}{2!} + \frac{x^4}{4!} + \frac{x^6}{6!} + \frac{x^8}{8!} + \cdots$$

for the hyperbolic cosine function, we get the following correspondences with the nomenclature used in Chapter 3.3.2:

$$A = a$$
$$B = 1$$
$$f(x,u) \equiv 0$$
$$f_u(x,u) \equiv 0$$
$$R = 1$$
$$N = 0$$
$$Q = q$$
$$\ell(x,u) = q\left(\frac{x^4}{4!} + \frac{x^6}{6!} + \frac{x^8}{8!} + \cdots \right)$$
$$\ell_u(x,u) \equiv 0 \ .$$

1st Approximation: LQ-Regulator

$$\dot{x}(t) = ax + u$$

$$J(u) = \int_0^\infty \left(\frac{1}{2}qx^2 + \frac{1}{2}u^2 \right) dt$$

$$u^{o(1)} = -Kx \ ,$$

where

$$K = a + \sqrt{a^2 + q}$$

is the positive solution of the Riccati equation

$$K^2 - 2aK - q = 0.$$

The resulting linear control system is described by the differential equation

$$\dot{x}(t) = [a - K]x(t) = A^o x(t) = -\sqrt{a^2 + q}\, x(t)$$

and has the cost-to-go function

$$\mathcal{J}^{(2)}(x) = \frac{1}{2}Kx^2 = \frac{1}{2}\left(a + \sqrt{a^2 + q}\right)x^2$$

with the derivative

$$\mathcal{J}_x^{[2]}(x) = Kx = \left(a + \sqrt{a^2 + q}\right)x\,.$$

2nd Approximation

From

$$0 = \mathcal{J}_x^{[3]} A^o x + \mathcal{J}_x^{[2]} f^{(2)} + \ell^{(3)}$$

we get

$$\mathcal{J}_x^{[3]} = 0\,.$$

Since $f_u(x, u) \equiv 0$, $\ell_u(x, u) \equiv 0$, $B = 1$, and $R = 1$, we obtain the following result for all $k \geq 2$:

$$u^{o(k)} = -\mathcal{J}_x^{[k+1]}\,.$$

Hence,

$$u^{o(2)} = -\mathcal{J}_x^{[3]} = 0\,.$$

3rd Approximation

$$0 = \mathcal{J}_x^{[4]} A^o x + \mathcal{J}_x^{[3]} B u^{o(2)} + \sum_{j=2}^{3} \mathcal{J}_x^{[5-j]} f^{(j)} + \frac{1}{2} u^{o(2)^{\mathrm{T}}} R u^{o(2)} + \ell^{(4)}$$

$$\mathcal{J}_x^{[4]} = \frac{qx^3}{4!\sqrt{a^2 + q}}$$

$$u^{o(3)} = -\mathcal{J}_x^{[4]} = -\frac{qx^3}{4!\sqrt{a^2 + q}}$$

4th Approximation

$$0 = \mathcal{J}_x^{[5]} A^\circ x + \sum_{j=2}^{3} \mathcal{J}_x^{[6-j]} B u^{o(j)} + \sum_{j=2}^{4} \mathcal{J}_x^{[6-j]} f^{(j)} + \sum_{j=2}^{2} u^{o(j)} R u^{o(5-j)} + \ell^{(5)}$$

$$\mathcal{J}_x^{[5]} = 0$$

$$u^{o(4)} = -\mathcal{J}_x^{[5]} = 0$$

5th Approximation

$$0 = \mathcal{J}_x^{[6]} A^\circ x + \sum_{j=2}^{4} \mathcal{J}_x^{[7-j]} B u^{o(j)} + \sum_{j=2}^{5} \mathcal{J}_x^{[7-j]} f^{(j)}$$

$$+ \sum_{j=2}^{2} u^{o(j)} R u^{o(6-j)} + \frac{1}{2} u^{o(3)} R u^{o(3)} + \ell^{(6)}$$

$$\mathcal{J}_x^{[6]} = \left(\frac{q}{6!} - \frac{q^2}{2(4!)^2(a^2+q)} \right) \frac{x^5}{\sqrt{a^2+q}}$$

$$u^{o(5)} = -\mathcal{J}_x^{[6]} = - \left(\frac{q}{6!} - \frac{q^2}{2(4!)^2(a^2+q)} \right) \frac{x^5}{\sqrt{a^2+q}}$$

6th Approximation

$$0 = \mathcal{J}_x^{[7]} A^\circ x + \sum_{j=2}^{5} \mathcal{J}_x^{[8-j]} B u^{o(j)} + \sum_{j=2}^{6} \mathcal{J}_x^{[8-j]} f^{(j)} + \sum_{j=2}^{3} u^{o(j)} R u^{o(7-j)} + \ell^{(7)}$$

$$\mathcal{J}_x^{[7]} = 0$$

$$u^{o(6)} = -\mathcal{J}_x^{[7]} = 0$$

7th Approximation

$$0 = \mathcal{J}_x^{[8]} A^\circ x + \sum_{j=2}^{6} \mathcal{J}_x^{[9-j]} B u^{o(j)} + \sum_{j=2}^{7} \mathcal{J}_x^{[9-j]} f^{(j)}$$

$$+ \sum_{j=2}^{3} u^{o(j)} R u^{o(8-j)} + \frac{1}{2} u^{o(4)} R u^{o(4)} + \ell^{(8)}$$

$$\mathcal{J}_x^{[8]} = \left(\frac{q}{8!} - \left(\frac{q}{6!} - \frac{q^2}{2(4!)^2(a^2+q)}\right)\frac{1}{4!(a^2+q)}\right)\frac{x^7}{\sqrt{a^2+q}}$$

$$u^{o(7)} = -\mathcal{J}_x^{[8]} = -\left(\frac{q}{8!} - \left(\frac{q}{6!} - \frac{q^2}{2(4!)^2(a^2+q)}\right)\frac{1}{4!(a^2+q)}\right)\frac{x^7}{\sqrt{a^2+q}}$$

and so on ...

Finally, we obtain the following nonlinear, approximatively optimal control

$$u^o(x) = u^{o(1)}(x) + u^{o(3)}(x) + u^{o(5)}(x) + u^{o(7)}(x) + \ldots .$$

Pragmatically, it can be approximated by the following equation:

$$u^o(x) \approx -(a+\sqrt{a^2+q})x - \frac{qx^3}{4!\sqrt{a^2+q}} - \frac{qx^5}{6!\sqrt{a^2+q}} - \frac{qx^7}{8!\sqrt{a^2+q}} - \ldots .$$

The characteristic of this approximated controller truncated after four terms is shown in Fig. 3.3.

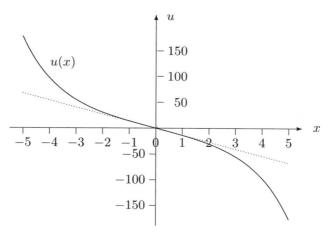

Fig. 3.3. Approximatively optimal controller for $a = 3$, $q = 100$.

3.3.4 LQQ Speed Control

The equation of motion for the velocity $v(t)$ of an aircraft in horizontal flight can be described by

$$m\dot{v}(t) = -\frac{1}{2}c_w A_r \rho v^2(t) + F(t) \ ,$$

where $F(t)$ is the horizontal thrust force generated by the jet engine, m is the mass of the aircraft, c_w is the aerodynamic drag coefficient, A_r is a reference cross section of the aircraft, and ρ is the density of the air.

The aircraft should fly at the constant speed v_0. For this, the nominal thrust

$$F_0 = \frac{1}{2}c_w A_r \rho v_0^2$$

is needed.

We want to augment the obvious open-loop control strategy $F(t) \equiv F_0$ with a feedback control such that the velocity $v(t)$ is controlled more precisely, should any discrepancy occur for whatever reason.

Introducing the state variable

$$x(t) = v(t) - v_0$$

and the correcting additive control variable

$$u(t) = \frac{1}{m}\big(F(t) - F_0\big)$$

the following nonlinear dynamics for the design of the feedback control are obtained:

$$\dot{x}(t) = a_1 x(t) + a_2 x^2(t) + u(t)$$

$$\text{with} \quad a_1 = -\frac{c_w A_r \rho v_0}{m}$$

$$\text{and} \quad a_2 = -\frac{c_w A_r \rho}{2m} \ .$$

For the design of the feedback controller, we choose the standard quadratic cost functional

$$J(u) = \frac{1}{2}\int_0^\infty \big(q x^2(t) + u^2(t)\big)\, dt \ .$$

Thus, we get the following correspondences with the nomenclature used in Chapter 3.3.2:

$$A = a_1$$
$$B = 1$$
$$f(x, u) = a_2 x^2$$
$$f_u(x, y) \equiv 0$$
$$f^{(1)}(x, u) = 2a_2 x$$
$$f^{(2)}(x, u) = 2a_2$$
$$f^{(3)}(x, u) = 0$$
$$Q = q$$
$$R = 1$$
$$\ell(x, u) \equiv 0 .$$

1st Approximation: LQ-Regulator

$$\dot{x}(t) = a_1 x + u$$
$$J(u) = \int_0^\infty \left(\frac{1}{2} q x^2 + \frac{1}{2} u^2 \right) dt$$
$$u^{o(1)} = -Kx ,$$

where

$$K = a_1 + \sqrt{a_1^2 + q}$$

is the positive solution of the Riccati equation

$$K^2 - 2a_1 K - q = 0 .$$

The resulting linear control system is described by the differential equation

$$\dot{x}(t) = [a_1 - K]x(t) = A^o x(t) = -\sqrt{a_1^2 + q}\, x(t)$$

and has the cost-to-go function

$$\mathcal{J}^{(2)}(x) = \frac{1}{2} K x^2 = \frac{1}{2} \left(a_1 + \sqrt{a_1^2 + q} \right) x^2$$

with the derivative

$$\mathcal{J}_x^{[2]}(x) = K x = \left(a_1 + \sqrt{a_1^2 + q} \right) x .$$

2ⁿᵈ Approximation

From

$$0 = \mathcal{J}_x^{[3]} A^\circ x + \mathcal{J}_x^{[2]} f^{(2)} + \ell^{(3)}$$

we get

$$\mathcal{J}_x^{[3]} = \frac{a_1 + \sqrt{a_1^2 + q}}{\sqrt{a_1^2 + q}} a_2 x^2 \ .$$

Since $f_u(x, u) \equiv 0$, $\ell_u(x, u) \equiv 0$, $B = 1$, and $R = 1$, we obtain the following result for all $k \geq 2$:

$$u^{o(k)} = -\mathcal{J}_x^{[k+1]} \ .$$

Hence,

$$u^{o(2)} = -\mathcal{J}_x^{[3]} = -\frac{a_1 + \sqrt{a_1^2 + q}}{\sqrt{a_1^2 + q}} a_2 x^2 \ .$$

Since the equation of motion is quadratic in x, the algorithm stops here. Therefore, the approximatively optimal control law is:

$$u(x) = u^{o(1)}(x) + u^{o(2)}(x) = -\left(a_1 + \sqrt{a_1^2 + q}\right) x - \frac{a_1 + \sqrt{a_1^2 + q}}{\sqrt{a_1^2 + q}} a_2 x^2 \ .$$

The characteristic of this approximated controller is shown in Fig. 3.4.

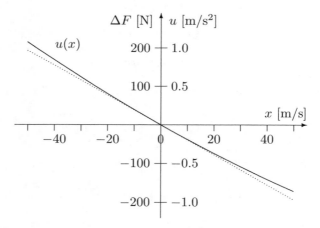

Fig. 3.4. Characteristic of the LQQ controller for $v_0 = 100\,\text{m/s}$ and $q = 0.001$ with $c_w = 0.05$, $A_r = 0.5\,\text{m}^2$, $\rho = 1.3\,\text{kg/m}^3$, and $m = 200\,\text{kg}$.

3.4 Exercises

1. Consider a bank account with the instantaneous wealth $x(t)$ and with the given initial wealth x_a at the given initial time $t_a = 0$. At any time, money can be withdrawn from the account at the rate $u(t) \geq 0$. The bank account receives interest. Therefore, it is an unstable system (which, alas, is easily stabilizable in practice). Modeling the system in continuous-time, its differential equation is

$$\dot{x}(t) = ax(t) - u(t)$$
$$x(0) = x_a \ ,$$

where $a > 0$ and $x_a > 0$. The compromise between withdrawing a lot of money from the account and letting the wealth grow due to the interest payments over a fixed time interval $[0, t_b]$ is formulated via the cost functional or "utility function"

$$J(u) = \frac{\alpha}{\gamma} x(t_b)^\gamma + \int_0^{t_b} \frac{1}{\gamma} u(t)^\gamma \, dt$$

which we want to maximize using an optimal state feedback control law. Here, $\alpha > 0$ is a parameter by which we influence the compromise between being rich in the end and consuming a lot in the time interval $[0, t_b]$. Furthermore, $\gamma \in (0, 1)$ is a "style parameter" of the utility function. Of course, we must not overdraw the account at any time, i.e., $x(t) \geq 0$ for all $t \in [0, t_b]$. And we can only withdraw money from the account, but we cannot invest money into the bank account, because our salary is too low. Hence, $u(t) \geq 0$ for all $t \in [0, t_b]$.

This problem can be solved analytically.

2. Find a state feedback control law for the asymptotically stable first-order system

$$\dot{x}(t) = ax(t) + bu(t) \quad \text{with} \quad a < 0, \ b > 0$$

such that the cost functional

$$J = kx^2(t_b) + \int_0^{t_b} \left(qx^2(t) + \cosh(u(t)) - 1 \right) dt$$

is minimized, where $k > 0$, $q > 0$, and t_b is fixed.

3. For the nonlinear time-invariant system of first order

$$\dot{x}(t) = a(x(t)) + b(x(t))u(t)$$

find a time-invariant state feedback control law, such that the cost functional

$$J(u) = \int_0^\infty \left(g(x(t)) + ru^{2k}(t) \right) dt$$

is minimized.

Here, the functions $a(x)$, $b(x)$, and $g(x)$ are continuously differentiable. Furthermore, the following conditions are satisfied:

$a(0) = 0$

$\dfrac{da}{dx}(0) \neq 0$

$a(.)$: either monotonically increasing or monotonically decreasing

$b(x) > 0$ for all $x \in R$

$g(0) = 0$

$g(x)$: strictly convex for all $x \in R$

$g(x) \to \infty$ for $|x| \to \infty$

$r > 0$

k : positive integer .

4. Consider the following "expensive control" version of the problem presented in Exercise 3:

   For the nonlinear time-invariant system of first order

   $$\dot{x}(t) = a(x(t)) + b(x(t))u(t)$$

   find a time-invariant state feedback control law, such that the system is stabilized and such that the cost functional

   $$J(u) = \int_0^\infty u^{2k}(t)\, dt$$

   is minimized for every initial state $x(0) \in R$.

5. Consider the following optimal control problem of Type B.1 where the cost functional contains an additional discrete state penalty term $K_1(x(t_1))$ at the fixed time t_1 within the time interval $[t_a, t_b]$:

   Find a piecewise continuous control $u : [t_a, t_b] \to \Omega$, such that the dynamic system

   $$\dot{x}(t) = f(x(t), u(t))$$

   is transferred from the initial state

   $$x(t_a) = x_a$$

   to the target set S at the fixed final time,

   $$x(t_b) \in S \subseteq R^n ,$$

   and such that the cost functional

   $$J(u) = K(x(t_b)) + K_1(x(t_1)) + \int_{t_a}^{t_b} L(x(t), u(t))\, dt$$

is minimized.

Prove that the additional discrete state penalty term $K_1(x(t_1))$ leads to the additional necessary jump discontinuity of the costate at t_1 of the following form:

$$\lambda^o(t_1^-) = \lambda^o(t_1^+) + \nabla_x K_1(x^o(t_1)) \, .$$

6. Consider the following optimal control problem of Type B.1 where there is an additional state constraint $x(t_1) \in S_1 \subset R^n$ at the fixed time t_1 within the time interval $[t_a, t_b]$:

Find a piecewise continuous control $u : [t_a, t_b] \to \Omega$, such that the dynamic system

$$\dot{x}(t) = f(x(t), u(t))$$

is transferred from the initial state

$$x(t_a) = x_a$$

through the loophole[4] or across (or onto) the surface[5]

$$x(t_1) \in S_1 \subset R^n$$

to the target set S at the fixed final time,

$$x(t_b) \in S \subseteq R^n \, ,$$

and such that the cost functional

$$J(u) = K(x(t_b)) + \int_{t_a}^{t_b} L(x(t), u(t)) \, dt$$

is minimized.

Prove that the additional discrete state constraint at time t_1 leads to the additional necessary jump discontinuity of the costate at t_1 of the following form:

$$\lambda^o(t_1^-) = \lambda^o(t_1^+) + q_1^o$$

where q_1^o satisfies the transversality condition

$$q_1 \in T^*(S_1, x^o(t_1)) \, .$$

Note that this phenomenon plays a major role in differential game problems. The major issue in differential game problems is that the involved "surfaces" are not obvious at the outset.

---

[4] A loophole is described by an inequality constraint $g_1(x(t_1)) \le 0$.
[5] A surface is described by an equality constraint $g_1(x(t_1)) = 0$.

4 Differential Games

A differential game problem is a generalized optimal control problem which involves two players rather than only one. One player chooses the control $u(t) \in \Omega_u \subseteq R^{m_u}$ and tries to minimize his cost functional, while the other player chooses the control $v(t) \in \Omega_v \subseteq R^{m_v}$ and tries to maximize her cost functional. — A differential game problem is called a zero-sum differential game if the two cost functionals are identical.

The most intriguing differential games are pursuit-evasion games, such as the homicidal chauffeur game, which has been stated as Problem 12 in Chapter 1 on p. 15. For its solution, consult [21] and [28].

This introduction into differential games is very short. Its raison d'être here lies in the interesting connections between differential games and the H_∞ theory of robust linear control.

In most cases, solving a differential game problem is mathematically quite tricky. The notable exception is the LQ differential game which is solved in Chapter 4.2. Its connections to the H_∞ control problem are analyzed in Chapter 4.3. For more detailed expositions of these connections, see [4] and [17].

The reader who is interested in more fascinating differential game problems should consult the seminal works [21] and [9] as well as the very complete treatise [5].

4.1 Theory

Conceptually, extending the optimal control theory to the differential game theory is straightforward and does not offer any surprises (initially): In Pontryagin's Minimum Principle, the Hamiltonian function has to be globally minimized with respect to the control u. In the corresponding Nash-Pontryagin Minimax Principle, the Hamiltonian function must simultaneously be globally minimized with respect to u and globally maximized with respect to v.

The difficulty is: In a general problem statement, the Hamiltonian function will not have such a minimax solution. — Pictorially speaking, the chance that a differential game problem (with a quite general formulation) has a solution is about as high as the chance that a horseman riding his saddled horse in the (u, v) plane at random happens to ride precisely in the Eastern (or Western) direction all the time.

Therefore, in addition to the general statement of the differential game problem, we also consider a special problem statement with "variable separation". — Yes, in dressage competitions, horses do perform traverses. (Nobody knows whether they think of differential games while doing this part of the show.)

For simplicity, we concentrate on time-invariant problems with unbounded controls u and v and with an unspecified final state at the fixed final time t_b.

4.1.1 Problem Statement

General Problem Statement

Find piecewise continuous controls $u : [t_a, t_b] \to R^{m_u}$ and $v : [t_a, t_b] \to R^{m_v}$, such that the dynamic system

$$\dot{x}(t) = f(x(t), u(t), v(t))$$

is transferred from the given initial state

$$x(t_a) = x_a$$

to an arbitrary final state at the fixed final time t_b and such that the cost functional

$$J(u, v) = K(x(t_b)) + \int_{t_a}^{t_b} L(x(t), u(t), v(t))\, dt$$

is minimized with respect to $u(.)$ and maximized with respect to $v(.)$.

Subproblem 1: Both users must use open-loop controls:

$$u(t) = u(t, x_a, t_a), \quad v(t) = v(t, x_a, t_a) \ .$$

Subproblem 2: Both users must use closed-loop controls in the form:

$$u(t) = k_u(x(t), t), \quad v(t) = k_v(x(t), t) \ .$$

Special Problem Statement with Separation of Variables

The functions f and L in the general problem statement have the following properties:

$$f(x(t), u(t), v(t)) = f_1(x(t), u(t)) + f_2(x(t), v(t))$$

$$L(x(t), u(t), v(t)) = L_1(x(t), u(t)) + L_2(x(t), v(t)) \ .$$

Remarks:

1) As mentioned in Chapter 1.1.2, the functions f, K, and L are assumed to be at least once continuously differentiable with respect to all of their arguments.

2) Obviously, the special problem with variable separation has a reasonably good chance to have an optimal solution. Furthermore, the existence theorems for optimal control problems given in Chapter 2.7 carry over to differential game problems in a rather straightforward way.

3) In the differential game problem with variable separation, the distinction between Subproblem 1 and Subproblem 2 is no longer necessary. As in optimal control problems, optimal open-loop strategies are equivalent to optimal closed-loop strategies (at least in theory). — In other words, condition c of the Theorem in Chapter 4.1.2 is automatically satisfied.

4) Since the final state is free, the differential game problem is regular, i.e., $\lambda_0^o = 1$ in the Hamiltonian function H.

4.1.2 The Nash-Pontryagin Minimax Principle

Definition: Hamiltonian function $H : R^n \times R^{m_u} \times R^{m_v} \times R^n \to R$,

$$H(x(t), u(t), v(t), \lambda(t)) = L(x(t), u(t), v(t)) + \lambda^{\mathrm{T}}(t) f(x(t), u(t), v(t)) .$$

Theorem

If $u^o : [t_a, t_b] \to R^{m_u}$ and $v^o : [t_a, t_b] \to R^{m_v}$ are optimal controls, then the following conditions are satisfied:

a) $\dot{x}^o(t) = \nabla_\lambda H_{|o} = f(x^o(t), u^o(t), v^o(t))$

   $x^o(t_a) = x_a$

   $\dot{\lambda}^o(t) = -\nabla_x H_{|o}$

   $\qquad = -\nabla_x L(x^o(t), u^o(t), v^o(t)) - \left[\dfrac{\partial f}{\partial x}(x^o(t), u^o(t), v^o(t)) \right]^{\mathrm{T}} \lambda^o(t)$

   $\lambda^o(t_b) = \nabla_x K(x^o(t_b)) .$

b) For all $t \in [t_a, t_b]$, the Hamiltonian $H(x^o(t), u, v, \lambda^o(t))$ has a global saddle point with respect to $u \in R^{m_u}$ and $v \in R^{m_v}$, and the saddle is correctly aligned with the control axes, i.e.,

   $H(x^o(t), u^o(t), v^o(t), \lambda^o(t)) \leq H(x^o(t), u, v^o(t), \lambda^o(t))$ for all $u \in R^{m_u}$

   and

   $H(x^o(t), u^o(t), v^o(t), \lambda^o(t)) \geq H(x^o(t), u^o(t), v, \lambda^o(t))$ for all $v \in R^{m_v}$.

c) Furthermore, in the case of Subproblem 2:

When the state feedback law $v(t) = k_v(x(t), t)$ is applied, $u^o(.)$ is a globally minimizing control of the resulting optimal control problem of Type C.1 and, conversely, when the state feedback law $u(t) = k_u(x(t), t)$ is applied, $v^o(.)$ is a globally maximizing control of the resulting optimal control problem of Type C.1.

4.1.3 Proof

Proving the theorem proceeds in complete analogy to the proofs of Theorem C in Chapter 2.3.3 and Theorem A in Chapter 2.1.3.

The augmented cost functional is:

$$\overline{J} = K(x(t_b)) + \int_{t_a}^{t_b} \left[L(x, u, v) + \lambda(t)^T \{ f(x, u, v) - \dot{x} \} \right] dt + \lambda_a^T \{ x_a - x(t_a) \}$$

$$= K(x(t_b)) + \int_{t_a}^{t_b} \left[H - \lambda^T \dot{x} \right] dt + \lambda_a^T \{ x_a - x(t_a) \} \ ,$$

where $H = H(x, u, v, \lambda) = L(x, u, v) + \lambda^T f(x, u, v)$ is the Hamiltonian function.

According to the philosophy of the Lagrange multiplier method, the augmented cost functional $\overline{J}$ has to be extremized with respect to all of its mutually independent variables $x(t_a)$, λ_a, $x(t_b)$, and $u(t)$, $v(t)$ $x(t)$, and $\lambda(t)$ for all $t \in (t_a, t_b)$.

Suppose that we have found the optimal solution $x^o(t_a)$, λ_a^o, $x^o(t_b)$, and $u^o(t)$, $v^o(t)$, $x^o(t)$, and $\lambda^o(t)$ for all $t \in (t_a, t_b)$.

The following first differential $\delta \overline{J}$ of $\overline{J}(u^o)$ around the optimal solution is obtained:

$$\delta \overline{J} = \left[\left(\frac{\partial K}{\partial x} - \lambda^T \right) \delta x \right]_{t_b} + \delta \lambda_a^T \{ x_a - x(t_a) \} + \left(\lambda^T(t_a) - \lambda_a^T \right) \delta x(t_a)$$

$$+ \int_{t_a}^{t_b} \left[\left(\frac{\partial H}{\partial x} + \dot{\lambda}^T \right) \delta x + \frac{\partial H}{\partial u} \delta u + \frac{\partial H}{\partial v} \delta v + \left(\frac{\partial H}{\partial \lambda} - \dot{x}^T \right) \delta \lambda \right] dt \ .$$

Since we have postulated a saddle point of the augmented function at $\overline{J}(u^o)$, this first differential must satisfy the following equality and inequalities

$$\delta \overline{J} \begin{cases} = 0 & \text{for all } \delta x, \ \delta \lambda, \text{ and } \delta \lambda_a \in R^n \\ \geq 0 & \text{for all } \delta u \in R^{m_u} \\ \leq 0 & \text{for all } \delta v \in R^{m_v} \ . \end{cases}$$

According to the philosophy of the Lagrange multiplier method, this equality and these inequalities must hold for arbitrary combinations of the mutually independent variations $\delta x(t)$, $\delta u(t)$, $\delta v(t)$, $\delta \lambda(t)$ at any time $t \in (t_a, t_b)$, and $\delta \lambda_a$, $\delta x(t_a)$, and $\delta x(t_b)$. Therefore, they must be satisfied for a few very specially chosen combinations of these variations as well, namely where only one single variation is nontrivial and all of the others vanish.

The consequence is that all of the factors multiplying a differential must vanish. — This completes the proof of the conditions a and b of the theorem.

Compared to Pontryagin's Minimum Principle, the condition c of the Nash-Pontryagin Minimax Principle is new. It should be fairly obvious because now, two *independent* players may use state feedback control. Therefore, if one player uses his optimal state feedback control law, the other player has to check whether Pontryagin's Minimum Principle is still satisfied for his (open-loop or closed-loop) control law. — This funny check only appears in differential game problems without separation of variables.

Notice that there is no condition for λ_a. In other words, the boundary condition $\lambda^o(t_a)$ of the optimal costate $\lambda^o(.)$ is free.

Remark: The calculus of variations only requires the local minimization of the Hamiltonian H with respect to the control u and a local maximization of H with respect to v. — In the theorem, the Hamiltonian is required to be globally minimized and maximized, respectively. Again, this restriction is justified in Chapter 2.2.1.

4.1.4 Hamilton-Jacobi-Isaacs Theory

In the Nash-Pontryagin Minimax Principle, we have expressed the necessary condition for H to have a Nash equilibrium or special type of saddle point with respect to (u, v) at (u^o, v^o) by the two inequalities

$$H(x^o, u^o, v, \lambda^o) \leq H(x^o, u^o, v^o, \lambda^o) \leq H(x^o, u, v^o, \lambda^o) \,.$$

In order to extend the Hamilton-Jacobi-Bellman theory in the area of optimal control to the Hamilton-Jacobi-Isaacs theory in the area of differential games, Nash's formulation of the necessary condition for a Nash equilibrium is more practical:

$$\min_u \max_v H(x^o, u, v, \lambda^o) = \max_v \min_u H(x^o, u, v, \lambda^o) = H(x^o, u^o, v^o, \lambda^o) \,,$$

i.e., it is not important whether H is first maximized with respect to v and then minimized with respect to u or vice versa. The result is the same in both cases.

Now, let us consider the following general time-invariant differential game problem with state feedback:

Find two state feedback control laws $u(x) : R^n \to R^{m_u}$ and $v : R^n \to R^{m_v}$, such that the dynamic system

$$\dot{x}(t) = f(x(t), u(t), v(t))$$

is transferred from the given initial state

$$x(t_a) = x_a$$

to an arbitrary final state at the fixed final time t_b and such that the cost functional

$$J(u, v) = K(x(t_b)) + \int_{t_a}^{t_b} L(x(t), u(t), v(t)) \, dt$$

is minimized with respect to $u(.)$ and maximized with respect to $v(.)$.

Let us assume that the Hamiltonian function

$$H = L(x, u, v) + \lambda^{\mathrm{T}} f(x, u, v)$$

has a unique Nash equilibrium for all $x \in R^n$ and all $\lambda \in R^n$. The corresponding H-minimizing and H-maximizing controls are denoted by $\widetilde{u}(x, \lambda)$ and $\widetilde{v}(x, \lambda)$, respectively. In this case, H is said to be "normal".

If the normality hypothesis is satisfied, the following sufficient condition for the optimality of a solution of the differential game problem is obtained.

Hamilton-Jacobi-Isaacs Theorem

If the cost-to-go function $\mathcal{J}(x, t)$ satisfies the boundary condition

$$\mathcal{J}(x, t_b) = K(x)$$

and the Hamilton-Jacobi-Isaacs partial differential equation

$$-\frac{\partial \mathcal{J}}{\partial t} = \min_u \max_v H(x, u, v, \nabla_x \mathcal{J}) = \max_v \min_u H(x, u, v, \nabla_x \mathcal{J})$$
$$= H(x, \widetilde{u}(x, \nabla_x \mathcal{J}), \widetilde{v}(x, \nabla_x \mathcal{J}), \nabla_x \mathcal{J})$$

for all $(x, t) \in R^n \times [t_a, t_b]$, then the state feedback control laws

$$u(x) = \widetilde{u}(x, \nabla_x \mathcal{J}) \quad \text{and} \quad v(x) = \widetilde{v}(x, \nabla_x \mathcal{J})$$

are globally optimal.

Proof: See [5].

4.2 The LQ Differential Game Problem

For convenience, the problem statement of the LQ differential game (Chapter 1.2, Problem 11, p. 15) is recapitulated here.

Find the piecewise continuous, unconstrained controls $u : [t_a, t_b] \to R^{m_u}$ and $v : [t_a, t_b] \to R^{m_v}$ such that the dynamic system

$$\dot{x}(t) = Ax(t) + B_1 u(t) + B_2 v(t)$$

is transferred from the given initial state

$$x(t_a) = x_a$$

to an arbitrary final state at the fixed final time t_b and such that the quadratic cost functional

$$J(u, v) = \frac{1}{2} x^{\mathrm{T}}(t_b) F x(t_b)$$
$$+ \frac{1}{2} \int_{t_a}^{t_b} \left(x^{\mathrm{T}}(t) Q x(t) + u^{\mathrm{T}}(t) u(t) - \gamma^2 v^{\mathrm{T}}(t) v(t) \right) dt ,$$

$$\text{with } F > 0 \text{ and } Q > 0 ,$$

is simultaneously minimized with respect to u and maximized with respect to v. Both players are allowed to use state feedback control. This is not relevant though, since the problem has separation of variables.

4.2.1 The LQ Differential Game Problem Solved with the Nash-Pontryagin Minimax Principle

The Hamiltonian function is

$$H = \frac{1}{2} x^{\mathrm{T}} Q x + \frac{1}{2} u^{\mathrm{T}} u - \frac{1}{2} \gamma^2 v^{\mathrm{T}} v + \lambda^{\mathrm{T}} Ax + \lambda^{\mathrm{T}} B_1 u + \lambda^{\mathrm{T}} B_2 v .$$

The following necessary conditions are obtained from the Nash-Pontryagin Minimax Principle:

$$\dot{x}^o = \nabla_\lambda H|_o = Ax^o + B_1 u^o + B_2 v^o$$
$$\dot{\lambda}^o = -\nabla_x H|_o = -Qx^o - A^{\mathrm{T}} \lambda^o$$
$$x^o(t_a) = x_a$$
$$\lambda^o(t_b) = Fx^o(t_b)$$
$$\nabla_u H|_o = 0 = u^o + B_1^{\mathrm{T}} \lambda^o$$
$$\nabla_v H|_o = 0 = -\gamma^2 v^o + B_2^{\mathrm{T}} \lambda^o .$$

Thus, the global minimax of the Hamiltonian function yields the following H-minimizing and H-maximizing control laws:

$$u^o(t) = -B_1^T \lambda^o(t)$$
$$v^o(t) = \frac{1}{\gamma^2} B_2^T \lambda^o(t) .$$

Plugging them into the differential equation for x results in the linear two-point boundary value problem

$$\dot{x}^o(t) = Ax^o(t) - B_1 B_1^T \lambda^o(t) + \frac{1}{\gamma^2} B_2 B_2^T \lambda^o(t)$$
$$\dot{\lambda}^o(t) = -Qx^o(t) - A^T \lambda^o(t)$$
$$x^o(t_a) = x_a$$
$$\lambda^o(t_b) = Fx^o(t_b) .$$

Converting the optimal controls from the open-loop to the closed-loop form proceeds in complete analogy to the case of the LQ regulator (see Chapter 2.3.4).

The two differential equations are homogeneous in $(x^o; \lambda^o)$ and at the final time t_b, the costate vector $\lambda(t_b)$ is a linear function of the final state vector $x^o(t_b)$. Therefore, the linear ansatz

$$\lambda^o(t) = K(t)x^o(t)$$

will work, where $K(t)$ is a suitable time-varying n by n matrix.

Differentiating this ansatz with respect to the time t, and considering the differential equations for the costate λ and the state x, and applying the ansatz in the differential equations leads to the following equation:

$$\dot{\lambda} = \dot{K}x + K\dot{x} = \dot{K}x + KAx - KB_1 B_1^T Kx + \frac{1}{\gamma^2} KB_2 B_2^T Kx$$
$$= -Qx - A^T Kx$$

or equivalently

$$\left(\dot{K} + A^T K + KA - KB_1 B_1^T K + \frac{1}{\gamma^2} KB_2 B_2^T K + Q \right) x \equiv 0 .$$

This equation must be satisfied at all times $t \in [t_a, t_b]$. Furthermore, we arrive at this equation, irrespective of the initial state x_a at hand, i.e., for all $x_a \in R^n$. Thus, the vector x in this equation may be an arbitrary vector in R^n. Therefore, the sum of matrices in the brackets must vanish.

The resulting optimal state-feedback control laws are

$$u^o(t) = -B_1^{\mathrm{T}} K(t) x^o(t) \quad \text{and}$$
$$v^o(t) = \frac{1}{\gamma^2} B_2^{\mathrm{T}} K(t) x^o(t) \ ,$$

where the symmetric, positive-definite n by n matrix $K(t)$ is the solution of the matrix Riccati differential equation

$$\dot{K}(t) = - A^{\mathrm{T}} K(t) - K(t) A - Q + K(t) \Big[B_1 B_1^{\mathrm{T}} - \frac{1}{\gamma^2} B_2 B_2^{\mathrm{T}} \Big] K(t)$$

with the boundary condition

$$K(t_b) = F$$

at the final time t_b.

Note: The parameter γ must be sufficiently large, such that $K(t)$ stays finite over the whole interval $[t_a, t_b]$.

4.2.2 The LQ Differential Game Problem Solved with the Hamilton-Jacobi-Isaacs Theory

Using the Hamiltonian function

$$H = \frac{1}{2} x^{\mathrm{T}} Q x + \frac{1}{2} u^{\mathrm{T}} u - \frac{1}{2} \gamma^2 v^{\mathrm{T}} v + \lambda^{\mathrm{T}} A x + \lambda^{\mathrm{T}} B_1 u + \lambda^{\mathrm{T}} B_2 v \ ,$$

the H-minimizing control

$$\widetilde{u}(x, \lambda) = -B_1^{\mathrm{T}} \lambda(t) \ ,$$

and the H-maximizing control

$$\widetilde{v}(x, \lambda) = \frac{1}{\gamma^2} B_2^{\mathrm{T}} \lambda(t) \ ,$$

the following symmetric form of the Hamilton-Jacobi-Isaacs partial differential equation can be obtained:

$$-\frac{\partial \mathcal{J}}{\partial t} = H\Big(x, \widetilde{u}(x, \nabla_x \mathcal{J}), \widetilde{v}(x, \nabla_x \mathcal{J}), \nabla_x \mathcal{J} \Big)$$

$$= \frac{1}{2} x^{\mathrm{T}} Q x - \frac{1}{2} (\nabla_x \mathcal{J})^{\mathrm{T}} B_1 B_1^{\mathrm{T}} \nabla_x \mathcal{J} + \frac{1}{2\gamma^2} (\nabla_x \mathcal{J})^{\mathrm{T}} B_2 B_2^{\mathrm{T}} \nabla_x \mathcal{J}$$

$$+ \frac{1}{2} (\nabla_x \mathcal{J})^{\mathrm{T}} A x + \frac{1}{2} x^{\mathrm{T}} A^{\mathrm{T}} \nabla_x \mathcal{J}$$

$$\mathcal{J}(x, t_b) = \frac{1}{2} x^{\mathrm{T}} F x \ .$$

Inspecting the boundary condition and the partial differential equation reveals that the following quadratic separation ansatz for the cost-to-go function will be successful:

$$J(x,t) = \frac{1}{2} x^{\mathrm{T}} K(t) x \text{ with } K(t_b) = F .$$

The symmetric, positive-definite n by n matrix function $K(.)$ remains to be found for $t \in [t_a, t_b)$.

The new, separated form of the Hamilton-Jacobi-Isaacs partial differential equation is

$$0 = \frac{1}{2} x^{\mathrm{T}} \Big(\dot{K}(t) + Q - K(t) B_1 B_1^{\mathrm{T}} K(t) + \frac{1}{\gamma^2} K(t) B_2 B_2^{\mathrm{T}} K(t)$$
$$+ K(t) A + A^{\mathrm{T}} K(t) \Big) x .$$

Since $x \in R^n$ is the independent state argument of the cost-to-go function $J(x,t)$, the partial differential equation is satisfied if and only if the matrix sum in the brackets vanishes.

Thus, finally, the following closed-loop optimal control laws are obtained for the LQ differential game problem:

$$u(x(t)) = - B_1^{\mathrm{T}} K(t) x(t)$$
$$v(x(t)) = \frac{1}{\gamma^2} B_2^{\mathrm{T}} K(x) x(t) ,$$

where the symmetric, positive-definite n by n matrix $K(t)$ is the solution of the matrix Riccati differential equation

$$\dot{K}(t) = - A^{\mathrm{T}} K(t) - K(t) A + K(t) B_1 B_1^{\mathrm{T}} K(t) - \frac{1}{\gamma^2} K(t) B_2 B_2^{\mathrm{T}} K(t) - Q$$

with the boundary condition

$$K(t_b) = F .$$

Note: The parameter γ must be sufficiently large, such that $K(t)$ stays finite over the whole interval $[t_a, t_b]$.

4.3 H_∞-Control via Differential Games

In this section, the so-called "full-information" H_∞-control problem for a linear time-invariant plant is investigated.

H_∞ Problem Statement

For the linear, time-invariant dynamic system

$$\dot{x}(t) = Ax(t) + B_1 w(t) + B_2 u(t)$$
$$z(t) = C_1 x(t) + D_{11} w(t) + D_{12} u(t)$$

find a linear, time-invariant controller of the form

$$u(t) = -Gx(t) \ ,$$

such that the H_∞-norm of the closed-loop control system with the disturbance input vector w to the design output z is bounded by a given value $\gamma > 0$, i.e.,

$$\sup_{\substack{w \in L_2(0,\infty) \\ \|w\|_2 = 1}} \frac{\|z\|_2}{\|w\|_2} = \|T_{zw}\|_\infty \leq \gamma \ .$$

Here, the norms $\|z\|_2$ and $\|w\|_2$ are defined as

$$\|z\|_2 = \sqrt{\int_0^\infty z^{\mathrm{T}}(t) z(t)\, dt}$$

$$\|w\|_2 = \sqrt{\int_0^\infty w^{\mathrm{T}}(t) w(t)\, dt} \ .$$

The following well-posedness conditions are needed:

1) $[A, B_2]$ stabilizable
2) $[A, C_1]$ detectable
3) $D_{12}^{\mathrm{T}} D_{12}$ invertible
4) D_{11} sufficiently small, i.e., $\bar{\sigma}(D_{11}) < \gamma$, where $\bar{\sigma}(D_{11})$ denotes the maximal singular value of D_{11}.

Differential Game Problem Statement

This H_∞ problem statement is equivalent to the following statement of a differential game problem:

For the linear, time-invariant dynamic system

$$\dot{x}(t) = Ax(t) + B_1 w(t) + B_2 u(t)$$
$$z(t) = C_1 x(t) + D_{11} w(t) + D_{12} u(t)$$

with an arbitrary initial state $x(0) = x_a$, find a linear, time-invariant controller
of the form

$$u(t) = -Gx(t) \ ,$$

such that the cost functional

$$J(u, w) = \frac{1}{2} \int_0^\infty \{z^{\mathrm{T}}(t)z(t) - \gamma^2 w^{\mathrm{T}}(t)w(t)\} \, dt \ ,$$

is both minimized with respect to $u(.)$ and maximized with respect to $w(.)$.

Analysis of the Problem

For the analysis of this infinite horizon differential game problem, the following substitutions are useful:

$$i = \begin{bmatrix} w \\ u \end{bmatrix}$$

$$B = [\, B_1 \quad B_2 \,]$$

$$D_{1\bullet} = [\, D_{11} \quad D_{12} \,]$$

$$\bar{R} = \begin{bmatrix} D_{11}^{\mathrm{T}} D_{11} - \gamma^2 I & D_{11}^{\mathrm{T}} D_{12} \\ D_{12}^{\mathrm{T}} D_{11} & D_{12}^{\mathrm{T}} D_{12} \end{bmatrix} \ .$$

The Hamiltonian function is:

$$\begin{aligned}
H &= \frac{1}{2} z^{\mathrm{T}} z - \frac{\gamma^2}{2} w^{\mathrm{T}} w + \lambda^{\mathrm{T}} \dot{x} \\
&= \frac{1}{2}(C_1 x + D_{1\bullet} i)^{\mathrm{T}}(C_1 x + D_{1\bullet} i) - \frac{\gamma^2}{2} w^{\mathrm{T}} w + \lambda^{\mathrm{T}} A x + \lambda^{\mathrm{T}} B i \\
&= \frac{1}{2} x^{\mathrm{T}} C_1^{\mathrm{T}} C_1 x + \frac{1}{2} x^{\mathrm{T}} C_1^{\mathrm{T}} D_{1\bullet} i + \frac{1}{2} i^{\mathrm{T}} D_{1\bullet}^{\mathrm{T}} C_1 x + \frac{1}{2} i^{\mathrm{T}} D_{1\bullet}^{\mathrm{T}} D_{1\bullet} i - \frac{\gamma^2}{2} w^{\mathrm{T}} w \\
&\quad + \frac{1}{2} \lambda^{\mathrm{T}} A x + \frac{1}{2} x^{\mathrm{T}} A^{\mathrm{T}} \lambda + \frac{1}{2} \lambda^{\mathrm{T}} B i + \frac{1}{2} i^{\mathrm{T}} B^{\mathrm{T}} \lambda \\
&= \frac{1}{2} x^{\mathrm{T}} C_1^{\mathrm{T}} C_1 x + \frac{1}{2} x^{\mathrm{T}} C_1^{\mathrm{T}} D_{1\bullet} i + \frac{1}{2} i^{\mathrm{T}} D_{1\bullet}^{\mathrm{T}} C_1 x + \frac{1}{2} i^{\mathrm{T}} \bar{R} i \\
&\quad + \frac{1}{2} \lambda^{\mathrm{T}} A x + \frac{1}{2} x^{\mathrm{T}} A^{\mathrm{T}} \lambda + \frac{1}{2} \lambda^{\mathrm{T}} B i + \frac{1}{2} i^{\mathrm{T}} B^{\mathrm{T}} \lambda \\
&= \frac{1}{2} \left[i + \bar{R}^{-1}(B^{\mathrm{T}} \lambda + D_{1\bullet}^{\mathrm{T}} C_1 x) \right] \bar{R} \left[i + \bar{R}^{-1}(B^{\mathrm{T}} \lambda + D_{1\bullet}^{\mathrm{T}} C_1 x) \right] \\
&\quad + \frac{1}{2} x^{\mathrm{T}} C_1^{\mathrm{T}} C_1 x + \frac{1}{2} \lambda^{\mathrm{T}} A x + \frac{1}{2} x^{\mathrm{T}} A^{\mathrm{T}} \lambda \\
&\quad - \frac{1}{2}(B^{\mathrm{T}} \lambda + D_{1\bullet}^{\mathrm{T}} C_1 x)^{\mathrm{T}} \bar{R}^{-1}(B^{\mathrm{T}} \lambda + D_{1\bullet}^{\mathrm{T}} C_1 x) \ .
\end{aligned}$$

By assumption, the top left element $D_{11}^{\mathrm{T}}D_{11} - \gamma^2 I$ of $\bar{R}$ is negative-definite and its bottom right element $D_{12}^{\mathrm{T}}D_{12}$ is positive-definite. Therefore, the Hamiltonian has a Nash equilibrium at

$$i = \begin{bmatrix} w \\ u \end{bmatrix} = -\bar{R}^{-1}(B^{\mathrm{T}}\lambda + D_{1\bullet}^{\mathrm{T}}C_1 x) \ ,$$

where it attains the value

$$H(i) = \frac{1}{2}x^{\mathrm{T}}C_1^{\mathrm{T}}C_1 x + \frac{1}{2}\lambda^{\mathrm{T}}Ax + \frac{1}{2}x^{\mathrm{T}}A^{\mathrm{T}}\lambda$$
$$- \frac{1}{2}(B^{\mathrm{T}}\lambda + D_{1\bullet}^{\mathrm{T}}C_1 x)^{\mathrm{T}}\bar{R}^{-1}(B^{\mathrm{T}}\lambda + D_{1\bullet}^{\mathrm{T}}C_1 x) \ .$$

Using the proven ansatz $\lambda(t) = Kx(t)$ with the symmetric n by n matrix K, we get

$$i = \begin{bmatrix} w \\ u \end{bmatrix} = -\bar{R}^{-1}(B^{\mathrm{T}}K + D_{1\bullet}^{\mathrm{T}}C_1)x \ ,$$

and

$$H(i) = \frac{1}{2}x^{\mathrm{T}}\Big(C_1^{\mathrm{T}}C_1 + KA + A^{\mathrm{T}}K$$
$$- (B^{\mathrm{T}}K + D_{1\bullet}^{\mathrm{T}}C_1)^{\mathrm{T}}\bar{R}^{-1}(B^{\mathrm{T}}K + D_{1\bullet}^{\mathrm{T}}C_1)\Big)x$$
$$= \frac{1}{2}x^{\mathrm{T}}\Big(C_1^{\mathrm{T}}C_1 - C_1^{\mathrm{T}}D_{1\bullet}\bar{R}^{-1}D_{1\bullet}^{\mathrm{T}}C_1 - KB\bar{R}^{-1}B^{\mathrm{T}}K$$
$$+ K(A - B\bar{R}^{-1}D_{1\bullet}C_1) + (A - B\bar{R}^{-1}D_{1\bullet}C_1)^{\mathrm{T}}K\Big)x \ .$$

As required for a time-invariant differential game problem with infinite terminal time, the Hamiltonian function is set to zero (for all $x \in R^n$) by choosing K such that it satisfies the algebraic matrix Riccati equation

$$0 = \bar{A}^{\mathrm{T}}K + K\bar{A} - K\bar{S}K + \bar{Q} \ ,$$

where, for convenience, the substitutions

$$\bar{A} = A - B\bar{R}^{-1}D_{1\bullet}C_1$$
$$\bar{S} = B\bar{R}^{-1}B^{\mathrm{T}}$$
$$\bar{Q} = C_1^{\mathrm{T}}C_1 - C_1^{\mathrm{T}}D_{1\bullet}\bar{R}^{-1}D_{1\bullet}^{\mathrm{T}}C_1$$

have been used. The matrix K must be chosen to be the unique stabilizing solution of the algebraic matrix Riccati equation such that the matrix $A - \bar{S}K$ is a stability matrix.

For more details, see [17].

Solutions to Exercises

Chapter 1

1. Define the new control variable $u \in [0,1]$ which has the time-invariant constraint set $\Omega = [0,1]$. With the new control variable $u(t)$, the engine torque $T(t)$ can be defined as

$$T(t) = T_{\min}(n(t)) + u(t)[T_{\max}(n(t)) - T_{\min}(n(t))] .$$

2. • Problem 1 is of Type A.2.
   • Problem 2 is of Type A.2.
   • Problem 3 is of Type A.1.
   • At first glance, Problem 4 appears to be of Type B.1. However, considering the special form $J = -x_3(t_b)$ of the cost functional, we see that it is of Type A.1 and that it has to be treated in the special way indicated in Chapter 2.1.6.
   • Problem 5 is of Type C.1.
   • Problem 6 is of Type D.1. But since exterminating the fish population prior to the fixed final time t_b cannot be optimal, this problem can be treated as a problem of Type B.1.
   • Problem 7 is of Type B.1.
   • Problem 8 is of Type B.1.
   • Problem 9 is of Type B.2 if it is analyzed in earth-fixed coordinates, but of Type A.2 if it is analyzed in body-fixed coordinates.
   • Problem 10 is of Type A.2.

3. We have found the optimal solution $x_1^o = 1$, $x_2^o = -1$, $\lambda^o = 2$, and the minimal value $f(x_1^o, x_2^o) = 2$. The straight line defined by the constraint $x_1 + x_2 = 0$ is tangent to the contour line $\{(x_1, x_2) \in R^2 \mid f(x_1, x_2) \equiv f(x_1^o, x_2^o) = 2\}$ at the optimal point (x_1^o, x_2^o). Hence, the two gradients $\nabla_x f(x_1^o, x_2^o)$ and $\nabla_x g(x_1^o, x_2^o)$ are colinear (i.e., proportional to each other). Only in this way, the necessary condition

$$\nabla_x F(x_1^o, x_2^o) = \nabla_x f(x_1^o, x_2^o) + \lambda^o \nabla_x g(x_1^o, x_2^o) = 0$$

can be satisfied.

4. We have found the optimal solution $x_1^o = 1$, $x_2^o = 1.5$, $\lambda_1^o = 0.5$, $\lambda_2^o = 3$, $\lambda_3^o = 0$, and the minimal value $f(x_1^o, x_2^o) = 3.25$.

   The three constraints $g_1(x_1, x_2) \leq 0$, $g_2(x_1, x_2) \leq 0$, and $g_3(x_1, x_2) \leq 0$ define an admissible set $S \in R^2$, over which the minimum of $f(x_1, x_2)$ is sought. The minimum (x_1^o, x_2^o) lies at the corner of S, where $g_1(x_1^o, x_2^o) = 0$ and $g_2(x_1^o, x_2^o) = 0$, while the third constraint is inactive: $g_3(x_1^o, x_2^o) < 0$.

   With this solution, the condition

$$\begin{aligned} \nabla_x F(x_1^o, x_2^o) =& \nabla_x f(x_1^o, x_2^o) + \lambda_1^o \nabla_x g_1(x_1^o, x_2^o) \\ &+ \lambda_2^o \nabla_x g_2(x_1^o, x_2^o) + \lambda_3^o \nabla_x g_3(x_1^o, x_2^o) = 0 \end{aligned}$$

   is satisfied.

5. The augmented function is

$$F(x, y, \lambda_0, \lambda_1, \lambda_2) = \lambda_0(2x^2 + 17xy + 3y^2) + \lambda_1(x - y - 2) + \lambda_2(x^2 + y^2 - 4).$$

   The solution is determined by the two constraints alone and therefore $\lambda_0^o = 0$. They admit the two solutions: $(x, y) = (2, 0)$ with $f(2, 0) = 8$ and $(x, y) = (0, 2)$ with $f(0, 2) = 12$. Thus, the constrained minimum of f is at $x^o = 2$, $y^o = 0$. — Although this is not very interesting anymore, the corresponding optimal values of the Lagrange multipliers would be $\lambda_1^o = 34$ and $\lambda_2^o = -10.5$.

Chapter 2

1. Since the system is stable, the equilibrium point (0,0) can be reached with the limited control within a finite time. Hence, there exists an optimal solution.

   Hamiltonian function:

$$H = 1 + \lambda_1 x_2 - \lambda_2 x_1 - \lambda_2 u .$$

   Costate differential equations:

$$\dot{\lambda}_1 = -\frac{\partial H}{\partial x_1} = \lambda_2$$

$$\dot{\lambda}_2 = -\frac{\partial H}{\partial x_2} = -\lambda_1$$

   The minimization of the Hamiltonian functions yields the control law

$$u = -\mathrm{sign}(\lambda_2) = \begin{cases} +1 & \text{for } \lambda_2 < 0 \\ 0 & \text{for } \lambda_2 = 0 \\ -1 & \text{for } \lambda_2 > 0 . \end{cases}$$

   The assignment of $u = 0$ for $\lambda_2 = 0$ is arbitrary. It has no consequences, since $\lambda_2(t)$ only crosses zero at discrete times.

The costate system $(\lambda_1(t), \lambda_2(t))$ is also a harmonic oscillator. Therefore, the optimal control changes between the values $+1$ and -1 periodically with the duration of the half-period of π units of time. In general, the duration of the last "leg" before the state vector reaches the origin of the state space will be shorter.

The optimal control law in the form of a state-vector feedback can easily be obtained by proceeding along the route taken in Chapter 2.1.4. For a constant value of the control u, the state evolves on a circle. For $u \equiv 1$, the circle is centered at $(+1, 0)$, for $u \equiv -1$, the circle is centered at $(-1, 0)$. In both cases, the state travels in the clock-wise direction.

Putting things together, the optimal state feedback control law depicted in Fig. S.1 is obtained.

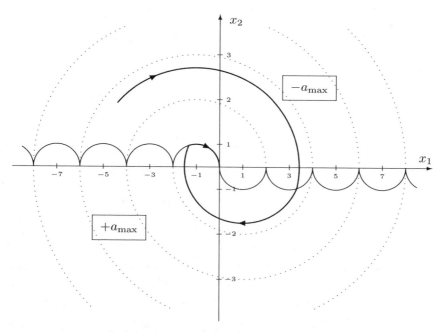

Fig. S.1. Optimal feedback control law for the time-optimal motion.

There is a switching curve consisting of the semi-circles centered at $(1, 0)$, $(3, 0)$, $(5, 0)$, ..., and at $(-1, 0)$, $(-3, 0)$, $(-5, 0)$, ... Below the switch curve, the optimal control is $u^o \equiv +1$, above the switch curve $u^o \equiv -1$. On the last leg of the trajectory, the optimal state vector travels along one of the two innermost semi-circles.

For a much more detailed analysis, the reader is referred to [2].

2. Since the control is unconstrained, there obviously exists an optimal non-singular solution. The Hamiltonian function is

$$H = u^2 + \lambda_1 x_2 + \lambda_2 x_2 + \lambda_2 u$$

and the necessary conditions for the optimality of a solution according to Pontryagin's Minimum Principle are:

a)
$$\dot{x}_1 = \partial H/\partial \lambda_1 = x_2$$

$$\dot{x}_2 = \partial H/\partial \lambda_2 = x_2 + u$$

$$\dot{\lambda}_1 = -\partial H/\partial x_1 = 0$$

$$\dot{\lambda}_2 = -\partial H/\partial x_2 = -\lambda_1 - \lambda_2$$

$$x_1(0) = 0$$

$$x_2(0) = 0$$

$$\begin{bmatrix} \lambda_1^o(t_b) \\ \lambda_2^o(t_b) \end{bmatrix} = \begin{bmatrix} q_1^o \\ q_2^o \end{bmatrix} \in T^*(S, x_1^o(t_b), x_2^o(t_b))$$

b)
$$\partial H/\partial u = 2u + \lambda_2 = 0 .$$

Thus, the open-loop optimal control law is $u^o(t) = -\frac{1}{2}\lambda_2^o(t)$.

Concerning $(x_1^o(t_b), x_2^o(t_b)) \in S$, we have to consider four cases:
- Case 1: $(x_1^o(t_b), x_2^o(t_b))$ lies in the interior of S:
  $x_1^o(t_b) > s_b$ and $x_2^o(t_b) < v_b$.
- Case 2: $(x_1^o(t_b), x_2^o(t_b))$ lies on the top boundary of S:
  $x_1^o(t_b) > s_b$ and $x_2^o(t_b) = v_b$.
- Case 3: $(x_1^o(t_b), x_2^o(t_b))$ lies on the left boundary of S:
  $x_1^o(t_b) = s_b$ and $x_2^o(t_b) < v_b$.
- Case 4: $(x_1^o(t_b), x_2^o(t_b))$ lies on the top left corner of S:
  $x_1^o(t_b) = s_b$ and $x_2^o(t_b) = v_b$.

In order to elucidate the discussion, let us call x_1 "position", x_2 "velocity", and u (the forced part of the) "acceleration".

It should be obvious that, with the given initial state, the cases 1 and 2 cannot be optimal. In both cases, a final state with the same final velocity could be reached at the left boundary of S or its top left corner, respectively, with a lower average velocity, i.e., at lower cost. Of course, these conjectures will be verified below.

Case 1: For a final state in the interior of S, the tangent cone of S is R^n (i.e., all directions in R^n) and the normal cone $T^*(S) = \{0\} \in R^2$. With $\lambda_1(t_b) = \lambda_2(t_b) = 0$, Pontryagin's necessary conditions cannot be satisfied because in this case, $\lambda_1(t) \equiv 0$, and $\lambda_2(t) \equiv 0$, and therefore, $u(t) \equiv 0$.

Case 2: On the top boundary, the normal cone $T^*(S)$ is described by $\lambda_1(t_b) = q_1 = 0$ and $\lambda_2(t_b) = q_2 > 0$. (We have already ruled out $q_2 = 0$.)

In this case, $\lambda_1(t) \equiv 0$, $\lambda_2(t) = q_2 e^{t_b - t} > 0$ at all times, and $u(t) = -\frac{1}{2}\lambda_2(t) < 0$ at all times. Hence, Pontryagin's necessary conditions cannot be satisfied. Of course, for other initial states with $x_2(0) > v_b$ and $x_1(0)$ sufficiently large, landing on the top boundary of S at the final time t_b is optimal.

Case 3: On the left boundary, the normal cone $T^*(S)$ is described by $\lambda_1(t_b) = q_1 < 0$ and $\lambda_2(t_b) = q_2 = 0$. (We have already ruled out $q_1 = 0$.) In this case, $\lambda_1(t) \equiv q_1 < 0$, $\lambda_2(t) = q_1 \left(e^{(t_b - t)} - 1 \right)$, and $u(t) = -\frac{1}{2}\lambda_2(t)$. The parameter $q_1 < 0$ has to be found such that the final position $x_1(t_b) = s_b$ is reached. This problem always has a solution. However, we have to investigate whether the corresponding final velocity satisfies the condition $x_2(t_b) \leq v_b$. If so, we have found the optimal solution, and the optimal control $u(t)$ is positive at all times $t < t_b$, but vanishes at the final time t_b. — If not, Case 4 below applies.

Case 4: In this last case, where the analysis of Case 3 yields a final velocity $x_2(t_b) > v_b$, the normal cone $T^*(S)$ is described by $\lambda_1(t_b) = q_1 < 0$ and $\lambda_2(t_b) > 0$. In this case, $\lambda_1(t) \equiv q_1 < 0$, $\lambda_2(t) = (q_1 + q_2)e^{(t_b - t)} - q_1$, and $u(t) = -\frac{1}{2}\lambda_2(t)$. The two parameters q_1 and q_2 have to be determined such that the conditions $x_1(t_b) = s_b$ and $x_2(t_b) = v_b$ are satisfied. There exists a unique solution. The details of this analysis are left to the reader. The major feature of this solution is that the control $u(t)$ will be positive in the initial phase and negative in the final phase.

3. In order to have a most interesting problem, let us assume that the specified final state (s_b, v_b) lies in the interior of the set $W(t_b) \subset R^2$ of all states which are reachable at the fixed final time t_b. This implies $\lambda_0^o = 1$.

Hamiltonian function: $H = u + \lambda_1 x_2 - \lambda_2 x_2^2 + \lambda_2 u = h(\lambda_2)u + \lambda_1 x_2 - \lambda_2 x_2^2$ with the switching function $h(\lambda_2) = 1 + \lambda_2$.

Pontryagin's necessary conditions for optimality:
a) Differential equations and boundary conditions:

$$\dot{x}_1^o = x_2^o$$
$$\dot{x}_2^o = -x_2^{o2} + u^o$$
$$\dot{\lambda}_1^o = 0$$
$$\dot{\lambda}_2^o = -\lambda_1^o + 2x_2^o \lambda_2^o$$
$$x_1^o(0) = 0$$
$$x_2^o(0) = v_a$$
$$x_1^o(t_b) = s_b$$
$$x_2^o(t_b) = v_b$$

b) Minimization of the Hamiltonian function:

$$h(\lambda_2^o(t))\, u^o(t) \leq h(\lambda_2^o(t))\, u$$

for all $u \in [0, 1]$ and all $t \in [0, t_b]$.

Preliminary open-loop control law:

$$u^o(t) \begin{cases} = 0 & \text{for } h(\lambda_2^o(t)) > 0 \\ = 1 & \text{for } h(\lambda_2^o(t)) < 0 \\ \in [0,1] & \text{for } h(\lambda_2^o(t)) = 0 \ . \end{cases}$$

Analysis of a potential singular arc:

$$h \equiv 0 = 1 + \lambda_2$$
$$\dot{h} \equiv 0 = \dot{\lambda}_2 = -\lambda_1 + 2x_2\lambda_2$$
$$\ddot{h} \equiv 0 = -\dot{\lambda}_1 + 2\dot{x}_2\lambda_2 + 2x_2\dot{\lambda}_2 = 2(u - x_2^2)\lambda_2 + 2x_2\dot{h} \ .$$

Hence, the optimal singular arc is characterized as follows:

$$u^o = x^{o2} \leq 1 \qquad \text{constant}$$
$$\lambda_2^o = -1 \qquad \text{constant}$$
$$\lambda_1^o = -2x_2^o \qquad \text{constant.}$$

The reader is invited to sketch all of the possible scenarios in the phase plane (x_1, x_2) for the cases $v_b > v_a$, $v_b = v_a$, and $v_b < v_a$.

4. Hamiltonian function:

$$H = \frac{1}{2}[y_d - Cx]^\mathrm{T} Q_y[y_d - Cx] + u^\mathrm{T} Ru + \lambda^\mathrm{T} Ax + \lambda^\mathrm{T} Bu \ .$$

The following control minimizes the Hamiltonian function:

$$u(t) = -R^{-1}B^\mathrm{T}\lambda \ .$$

Plugging this control law into the differential equations for the state x and the costate λ yields the following linear inhomogeneous two-point boundary value problem:

$$\dot{x} = Ax - BR^{-1}B^\mathrm{T}\lambda$$
$$\dot{\lambda} = -C^\mathrm{T}Q_yCx - A^\mathrm{T}\lambda + C^\mathrm{T}Q_y y_d$$
$$x(t_a) = x_a$$
$$\lambda(t_b) = C^\mathrm{T}F_yCx(t_b) - C^\mathrm{T}F_y y_d(t_b) \ .$$

In order to convert the resulting open-loop optimal control law into a closed-loop control law using state feedback, the following ansatz is suitable:

$$\lambda(t) = K(t)x(t) - w(t) \ ,$$

where the n by n matrix $K(.)$ and the n-vector $w(.)$ remain to be found. Plugging this ansatz into the two-point boundary value problem leads to the following differential equations for $K(.)$ and $w(.)$, which need to be solved in advance:

$$\dot{K} = -A^{\mathrm{T}}K - KA + KBR^{-1}B^{\mathrm{T}}K - C^{\mathrm{T}}Q_yC$$
$$\dot{w} = -[A - BR^{-1}BK]^{\mathrm{T}}w - C^{\mathrm{T}}Q_yy_d$$
$$K(t_b) = C^{\mathrm{T}}(t_b)F_yC(t_b)$$
$$w(t_b) = C^{\mathrm{T}}(t_b)F_yy_d(t_b) \ .$$

Thus, the optimal combination of a state feedback control and a feedforward control considering the future of $y_d(.)$ is:

$$u(t) = -R^{-1}(t)B^{\mathrm{T}}(t)K(t)x(t) + R^{-1}(t)B^{\mathrm{T}}(t)w(t) \ .$$

For more details, consult [2] and [16].

5. First, consider the following homogeneous matrix differential equation:

$$\dot{\Sigma}(t) = A^*(t)\Sigma(t) + \Sigma(t)A^{*\mathrm{T}}(t)$$
$$\Sigma(t_a) = \Sigma_a \ .$$

Its closed-form solution is

$$\Sigma(t) = \Phi(t, t_a)\Sigma_a\Phi^{\mathrm{T}}(t, t_a) \ ,$$

where $\Phi(t, t_a)$ is the transition matrix corresponding to the dynamic matrix $A^*(t)$. For arbitrary times t and τ, this transition matrix satisfies the following equations:

$$\Phi(t, t) = I$$
$$\frac{d}{dt}\Phi(t, \tau) = A^*(t)\Phi(t, \tau)$$
$$\frac{d}{d\tau}\Phi(t, \tau) = -\Phi(t, \tau)A^*(\tau) \ .$$

The closed-form solution for $\Sigma(t)$ can also be written in the operator notation

$$\Sigma(t) = \Psi(t, t_a)\Sigma_a \ ,$$

where the operator $\Psi(.,.)$ is defined (in the "maps to" form) by

$$\Psi(t, t_a) : \Sigma_a \mapsto \Phi(t, t_a)\Sigma_a\Phi^{\mathrm{T}}(t, t_a) \ .$$

Obviously, $\Psi(t, t_a)$ is a positive operator because it maps every positive-semidefinite matrix Σ_a to a positive-semidefinite matrix $\Sigma(t)$.

Sticking with the "maps to" notation and using differential calculus, we find the following useful results:

$$\Psi(t, \tau) : \Sigma_\tau \mapsto \Phi(t, \tau)\Sigma_\tau\Phi^{\mathrm{T}}(t, \tau)$$
$$\frac{d}{dt}\Psi(t, \tau) : \Sigma_\tau \mapsto A^*(t)\Phi(t, \tau)\Sigma_\tau\Phi^{\mathrm{T}}(t, \tau) + \Phi(t, \tau)\Sigma_\tau\Phi^{\mathrm{T}}(t, \tau)A^{*\mathrm{T}}(t)$$
$$\frac{d}{d\tau}\Psi(t, \tau) : \Sigma_\tau \mapsto -\Phi(t, \tau)A^*(\tau)\Sigma_\tau\Phi^{\mathrm{T}}(t, \tau) - \Phi(t, \tau)\Sigma_\tau A^{*\mathrm{T}}(\tau)\Phi^{\mathrm{T}}(t, \tau) \ .$$

Reverting now to operator notation, we have found the following results:

$$\Psi(t,t) = I$$

$$\frac{d}{d\tau}\Psi(t,\tau) = -\Psi(t,\tau)UA^*(\tau)$$

where the operator U has been defined on p. 71.

Considering this result for $A^* = A - PC$ and comparing it with the equations describing the costate operator λ^o (in Chapter 2.8.4) establishes that $\lambda^o(t)$ is a positive operator at all times $t \in [t_a, t_b]$, because $\Psi(.,.)$ is a positive operator irrespective of the underlying matrix A^*.

In other words, infimizing the Hamiltonian H is equivalent to infimizing $\dot{\Sigma}$. Of course, we have already exploited the necessary condition $\partial\dot{\Sigma}/\partial P = 0$, because the Hamiltonian is of the form $H = \lambda\,\dot{\Sigma}(P)$.

The fact that we have truly infimized the Hamiltonian and $\dot{\Sigma}$ with respect to the observer gain matrix P is easily established by expressing $\dot{\Sigma}$ in the form of a "complete square" as follows:

$$\dot{\Sigma} = A\Sigma - PC\Sigma + \Sigma A^{\mathrm{T}} - \Sigma C^{\mathrm{T}}P^{\mathrm{T}} + BQB^{\mathrm{T}} + PRP^{\mathrm{T}}$$
$$= A\Sigma + \Sigma A^{\mathrm{T}} + BQB^{\mathrm{T}} - \Sigma C^{\mathrm{T}}R^{-1}C\Sigma$$
$$+ [P - \Sigma C^{\mathrm{T}}R^{-1}]R[P - \Sigma C^{\mathrm{T}}R^{-1}]^{\mathrm{T}} \ .$$

The last term vanishes for the optimal choice $P^o = \Sigma C^{\mathrm{T}}R^{-1}$; otherwise it is positive-semidefinite. — This completes the proof.

Chapter 3

1. Hamiltonian function: $H = \dfrac{1}{\gamma}u^\gamma + \lambda ax - \lambda u$

   Maximizing the Hamiltonian:

$$\frac{\partial H}{\partial u} = u^{\gamma-1} - \lambda = 0$$

$$\frac{\partial^2 H}{\partial u^2} = (\gamma-1)u^{\gamma-2} < 0 \ .$$

Since $0 < \gamma < 1$ and $u \geq 0$, the H-maximizing control is

$$u = \lambda^{\frac{1}{\gamma-1}} \geq 0 \ .$$

In the Hamilton-Jacobi-Bellman partial differential equation $\frac{\partial J}{\partial t} + H = 0$ for the optimal cost-to-go function $J(x,t)$, λ has to be replaced by $\frac{\partial J}{\partial x}$ and u by the H-maximizing control

$$u = \left(\frac{\partial J}{\partial x}\right)^{\frac{1}{\gamma-1}} \ .$$

Thus, the following partial differential equation is obtained:

$$\frac{\partial \mathcal{J}}{\partial t} + \frac{\partial \mathcal{J}}{\partial x} ax + \left(\frac{1}{\gamma} - 1\right)\left(\frac{\partial \mathcal{J}}{\partial x}\right)^{\frac{\gamma}{\gamma-1}} = 0 \; .$$

According to the final state penalty term of the cost functional, its boundary condition at the final time t_b is

$$\mathcal{J}(x, t_b) = \frac{\alpha}{\gamma} x^\gamma \; .$$

Inspecting the boundary condition and the partial differential equation reveals that the following separation ansatz for the cost-to-go function will be successful:

$$\mathcal{J}(x, t) = \frac{1}{\gamma} x^\gamma G(t) \quad \text{with } G(t_b) = \alpha \; .$$

The function $G(t)$ for $t \in [t_a, t_b)$ remains to be determined.

Using

$$\frac{\partial \mathcal{J}}{\partial t} = \frac{1}{\gamma} x^\gamma \dot{G} \quad \text{and} \quad \frac{\partial \mathcal{J}}{\partial x} = x^{\gamma-1} G$$

the following form of the Hamilton-Jacobi-Bellman partial differential equation is obtained:

$$\frac{1}{\gamma} x^\gamma \left[\dot{G} + \gamma G a - (\gamma-1) G^{\frac{\gamma}{\gamma-1}} \right] = 0 \; .$$

Therefore, the function G has to satisfy the ordinary differential equation

$$\dot{G}(t) + \gamma a G(t) - (\gamma-1) G^{\frac{\gamma}{\gamma-1}}(t) = 0$$

with the boundary condition

$$G(t_b) = \alpha \; .$$

This differential equation is of the Bernoulli type. It can be transformed into a linear ordinary differential by introducing the substitution

$$Z(t) = G^{-\frac{1}{\gamma-1}}(t) \; .$$

The resulting differential equations is

$$\dot{Z}(t) = \frac{\gamma}{\gamma - 1} a Z(t) - 1$$

$$Z(t_b) = \alpha^{-\frac{1}{\gamma-1}} \; .$$

With the simplifying substitutions

$$A = \frac{\gamma}{\gamma - 1} a \quad \text{and} \quad Z_b = \alpha^{-\frac{1}{\gamma-1}} \ ,$$

the closed-form solution for $Z(t)$ is

$$Z(t) = \left[\left(Z_b - \frac{1}{A} \right) e^{-At_b} + \frac{1}{A} \right] e^{At} + \frac{1}{A} \left(1 - e^{At} \right) \ .$$

Finally, the following optimal state feedback control law results:

$$u(x(t)) = G^{\frac{1}{\gamma-1}}(t) x(t) = \frac{x(t)}{Z(t)} \ .$$

Since the inhomogeneous part in the first-order differential equation for $Z(t)$ is negative and the final value $Z(t_b)$ is positive, the solution $Z(t)$ is positive for all $t \in [0, t_b]$. In other words, we are always consuming, at a lower rate in the beginning and at a higher rate towards the end.

2. Hamiltonian function:

$$H = qx^2 + \cosh(u) - 1 + \lambda ax + \lambda bu \ .$$

Minimizing the Hamiltonian function:

$$\frac{\partial H}{\partial u} = \sinh(u) + b\lambda = 0 \ .$$

H-minimizing control:

$$u = \text{arsinh}(-b\lambda) = -\text{arsinh}(b\lambda) \ .$$

Hamilton-Jacobi-Bellman partial differential equation:

$$0 = \frac{\partial \mathcal{J}}{\partial t} + H$$

$$= \frac{\partial \mathcal{J}}{\partial t} + qx^2 + \cosh \left(\text{arsinh} \left(b \frac{\partial \mathcal{J}}{\partial x} \right) \right) - 1 + \frac{\partial \mathcal{J}}{\partial x} ax - b \frac{\partial \mathcal{J}}{\partial x} \text{arsinh} \left(b \frac{\partial \mathcal{J}}{\partial x} \right)$$

with the boundary condition

$$\mathcal{J}(x, t_b) = kx^2 \ .$$

Maybe this looks rather frightening, but this partial differential equation ought to be amenable to numerical integration.

3. Hamiltonian function: $H = g(x) + ru^{2k} + \lambda a(x) + \lambda b(x)u$.

The time-invariant state feedback controller is determined by the following two equations:

$$H = g(x) + ru^{2k} + \lambda a(x) + \lambda b(x)u \equiv 0 \qquad \text{(HJB)}$$
$$\nabla_u H = 2kru^{2k-1} + \lambda b(x) = 0 . \qquad (H_{\min})$$

The costate vector can be eliminated by solving $(H_{\min})$ for λ and plugging the result into (HJB). The result is

$$(2k-1)ru^{2k} + 2kr\frac{a(x)}{b(x)}u^{2k-1} - g(x) = 0 .$$

Thus, for every value of x, the zeros of this fairly special polynomial have to be found. According to Descartes' rule, this polynomial has exactly one positive real and exactly one negative real zero for all $k \geq 1$ and all $x \neq 0$. One of them will be the correct solution, namely the one which stabilizes the nonlinear dynamic system. From Theorem 1 in Chapter 2.7, we can infer that the optimal solution is unique.

The final result can be written in the following way:

$$u(x) = \underset{\substack{u \in R \\ \text{stabilizing}}}{\arg} \left[(2k-1)ru^{2k} + 2kr\frac{a(x)}{b(x)}u^{2k-1} - g(x) = 0 \right] .$$

4. In the case of "expensive control" with $g(x) \equiv 0$, we have to find the stabilizing solution of the polynomial

$$(2k-1)u^{2k} + 2k\frac{a(x)}{b(x)}u^{2k-1} = u^{2k-1}\left[(2k-1)u + 2k\frac{a(x)}{b(x)} \right] = 0 .$$

For the unstable plant with $a(x)$ monotonically increasing, the stabilizing controller is

$$u = -\frac{2k}{2k-1}\frac{a(x)}{b(x)} ,$$

for the asymptotically stable plant with $a(x)$ monotonically decreasing, it is optimal to do nothing, i.e.,

$$u \equiv 0 .$$

5. Applying the principle of optimality around the time t_1 yields

$$\mathcal{J}(x, t_1^-) = \mathcal{J}(x, t_1^+) + K_1(x) .$$

Since the costate is the gradient of the optimal cost-to-go function, the claim

$$\lambda^o(t_1^-) = \lambda^o(t_1^+) + \nabla_x K_1(x^o(t_1))$$

is established.

For more details, see [13].

6. Applying the principle of optimality at the time t_1 results in the antecedent optimal control problem of Type B.1 with the final state penalty term $\mathcal{J}(x, t_1^+)$ and the target set $x(t_1) \in S_1$. According to Theorem B in Chapter 2.5, this leads to the claimed result

$$\lambda^o(t_1^-) = \lambda^o(t_1^+) + q_1^0$$

where q_1^o lies in the normal cone $T^*(S_1, x^o(t_1))$ of the tangent cone $T(S_1, x^o(t_1))$ of the constraint set S_1 at $x^o(t_1)$.

For more details, see [10, Chapter 3.5].

References

1. B. D. O. Anderson, J. B. Moore, *Optimal Control: Linear Quadratic Methods*, Prentice-Hall, Englewood Cliffs, NJ, 1990.

2. M. Athans, P. L. Falb, *Optimal Control*, McGraw-Hill, New York, NY, 1966.

3. M. Athans, H. P. Geering, "Necessary and Sufficient Conditions for Differentiable Nonscalar-Valued Functions to Attain Extrema," *IEEE Transactions on Automatic Control, 18 (1973)*, pp. 132–139.

4. T. Başar, P. Bernhard, H^∞-*Optimal Control and Related Minimax Design Problems: A Dynamic Game Approach*, Birkhäuser, Boston, MA, 1991.

5. T. Başar, G. J. Olsder, *Dynamic Noncooperative Game Theory*, SIAM, Philadelphia, PA, 2nd ed., 1999.

6. R. Bellman, R. Kalaba, *Dynamic Programming and Modern Control Theory*, Academic Press, New York, NY, 1965.

7. D. P. Bertsekas, A. Nedić, A. E. Ozdaglar, *Convex Analysis and Optimization*, Athena Scientific, Nashua, NH, 2003.

8. D. P. Bertsekas, *Dynamic Programming and Optimal Control*, Athena Scientific, Nashua, NH, 3rd ed., vol. 1, 2005, vol. 2, 2007.

9. A. Blaquière, F. Gérard, G. Leitmann, *Quantitative and Qualitative Games*, Academic Press, New York, NY, 1969.

10. A. E. Bryson, Y.-C. Ho, *Applied Optimal Control*, Halsted Press, New York, NY, 1975.

11. A. E. Bryson, *Applied Linear Optimal Control*, Cambridge University Press, Cambridge, U.K., 2002.

12. H. P. Geering, M. Athans, "The Infimum Principle," *IEEE Transactions on Automatic Control, 19 (1974)*, pp. 485–494.

13. H. P. Geering, "Continuous-Time Optimal Control Theory for Cost Functionals Including Discrete State Penalty Terms," *IEEE Transactions on Automatic Control, 21 (1976)*, pp. 866–869.

14. H. P. Geering et al., *Optimierungsverfahren zur Lösung deterministischer regelungstechnischer Probleme*, Haupt, Bern, 1982.

15. H. P. Geering, L. Guzzella, S. A. R. Hepner, C. H. Onder, "Time-Optimal Motions of Robots in Assembly Tasks," *Transactions on Automatic Control, 31 (1986)*, pp. 512–518.

16. H. P. Geering, *Regelungstechnik*, 6th ed., Springer-Verlag, Berlin, 2003.

17. H. P. Geering, *Robuste Regelung*, Institut für Mess- und Regeltechnik, ETH, Zürich, 3rd ed., 2004.

18. B. S. Goh, "Optimal Control of a Fish Resource," *Malayan Scientist, 5 (1969/70)*, pp. 65–70.

19. B. S. Goh, G. Leitmann, T. L. Vincent, "Optimal Control of a Prey-Predator System," *Mathematical Biosciences, 19 (1974)*, pp. 263–286.

20. H. Halkin, "On the Necessary Condition for Optimal Control of Non-Linear Systems," *Journal d'Analyse Mathématique, 12 (1964)*, pp. 1–82.

21. R. Isaacs, *Differential Games: A Mathematical Theory with Applications to Warfare and Pursuit, Control and Optimization*, Wiley, New York, NY, 1965.

22. C. D. Johnson, "Singular Solutions in Problems of Optimal Control," in C. T. Leondes (ed.), *Advances in Control Systems, vol. 2*, pp. 209–267, Academic Press, New York, NY, 1965.

23. D. E. Kirk, *Optimal Control Theory: An Introduction*, Dover Publications, Mineola, NY, 2004.

24. R. E. Kopp, H. G. Moyer, "Necessary Conditions for Singular Extremals", *AIAA Journal, vol .3 (1965)*, pp. 1439–1444.

25. H. Kwakernaak, R. Sivan, *Linear Optimal Control Systems*, Wiley-Interscience, New York, NY, 1972.

26. E. B. Lee, L. Markus, *Foundations of Optimal Control Theory*, Wiley, New York, NY, 1967.

27. D. L. Lukes, "Optimal Regulation of Nonlinear Dynamical Systems," *SIAM Journal Control, 7 (1969)*, pp. 75–100.

28. A. W. Merz, *The Homicidal Chauffeur — a Differential Game*, Ph.D. Dissertation, Stanford University, Stanford, CA, 1971.

29. L. S. Pontryagin, V. G. Boltyanskii, R. V. Gamkrelidze, E. F. Mishchenko, *The Mathematical Theory of Optimal Processes*, (translated from the Russian), Interscience Publishers, New York, NY, 1962.

30. B. Z. Vulikh, *Introduction to the Theory of Partially Ordered Spaces*, (translated from the Russian), Wolters-Noordhoff Scientific Publications, Groningen, 1967.

31. L. A. Zadeh, "Optimality and Non-Scalar-Valued Performance Criteria," *IEEE Transactions on Automatic Control, 8 (1963)*, pp. 59–60.

Index